AF595045

Recent Developments and Applications in Tissue Mechanics and Tissue Engineering

Recent Developments and Applications in Tissue Mechanics and Tissue Engineering

Editor

Federica Boschetti

MDPI • Basel • Beijing • Wuhan • Barcelona • Belgrade • Manchester • Tokyo • Cluj • Tianjin

Editor
Federica Boschetti
Department of Chemistry, Materials and Chemical Engineering "Giulio Natta"
Politecnico di Milano
Italy

Editorial Office
MDPI
St. Alban-Anlage 66
4052 Basel, Switzerland

This is a reprint of articles from the Special Issue published online in the open access journal *Applied Sciences* (ISSN 2076-3417) (available at: https://www.mdpi.com/journal/applsci/special_issues/tissue_mechanics).

For citation purposes, cite each article independently as indicated on the article page online and as indicated below:

LastName, A.A.; LastName, B.B.; LastName, C.C. Article Title. *Journal Name* **Year**, *Volume Number*, Page Range.

ISBN 978-3-0365-5317-7 (Hbk)
ISBN 978-3-0365-5318-4 (PDF)

Cover image courtesy of Federica Boschetti
A porcine cornea under pressurization test.

Contents

About the Editor

Federica Boschetti

Federica Boschetti was born in 1965 in Milan (Italy). She received her Master's degree (M.Eng.) in Electronical Engineering in 1990 and her Ph.D. in Bioengineering in 1994 from Politecnico di Milano, Milan, Italy. From 1994 to 2005 she was a postdoc at Politecnico di Milano as well as at Ospedale S.Raffaele, Milan and Northwestern University Evanston, IL, USA. She was Assistant Professor at Politecnico di Milano from 2005 to 2013. Presently, is an Associate Professor of Industrial Bioengineering (since 2013) at the Department of Chemistry, Materials and Chemical Engineering, Giulio Natta of Politecnico di Milano and the co-founder of MgShell, an offshoot of Politecnico di Milano that is dedicated to the design, development, production, characterization, and commercialization of biomedical devices. Most of Federica Boschetti's research deals with experimental and computational biomechanics. After early research experience in mathematical modelling (numerical modelling of lumped parameters models, transport, and fluid dynamics) and in vitro tests with artificial lungs, in the last thirteen years, Federica Boschetti has been re-orienting her research interests towards tissue mechanics and tissue engineering; in vitro testing and design; and microfluid dynamics. The combined numerical–experimental approach is the common feature of all of her research activities. As of 2015, Federica Boschetti's most recent activities are directed towards drug delivery and device fabrication for the ocular environment in particular.

MDPI

Editorial

Tissue Mechanics and Tissue Engineering

Federica Boschetti

Dipartimento di Chimica, Materiali e Ingegneria Chimica, Politecnico di Milano, 20133 Milano, Italy; federica.boschetti@polimi.it

1. Introduction

Tissue engineering (TE) combines scaffolds, cells, and chemical and physical cues to replace biological tissues. Several disciplines, such as biology, chemistry, materials science, mathematics, and most branches of engineering, support this goal while improving the quality of the reconstructed tissues. Scaffolds are designed to mimic the native extracellular matrix, thus allowing cell growth and differentiation while avoiding adverse reactions [1,2]. Autologous cells or co-cultures of donor cells either encapsulated or seeded on scaffolds are deeply studied in response to chemical and physical stimuli with an effort to make them survive, grow, and differentiate [3].

As far as concerns with physical stimulation, scaffolds seeded with cells are subjected to mechanical stimuli through a bioreactor to obtain the desired functional tissue in vitro. Moreover, the scaffold itself must have mechanical properties similar to the ones of the native tissue. Tissue mechanics (TM) study the response of living tissues to applied loads, which is important for understanding the structural role of tissues and how cells respond to mechanical stimuli. Hence, mechanical properties used to quantify the load response are usually measured for native tissues and serve as reference properties for regenerated tissues. In other words, obtaining the normal living tissue mechanical properties is one of the goals of researchers for tissue regeneration. Diseases often change tissue mechanical properties, and therefore, their measurement may help detect and follow the evolution of pathologies, thus improving the clinical outcome of the patients [4,5].

2. Review of Special Issue Contents

Contributions to this Special Issue focus on different aspects of Tissue Mechanics and Tissue Engineering, giving valuable examples of applied research in the field. Two papers face different applications of tissue regeneration: the skin [1] and the retina [2]. One paper [3] deals with cell encapsulation in biomaterials for protection against immune reactions. Two papers underline that tissue mechanics play a key role in understanding tissue functions, the former focusing on the role of biomechanics related to the prediction of aortic aneurysm rupture [4], whereas the latter on the role of biomechanics for early diagnosis of glaucoma [5].

Eisler et al. [1] investigated the behavior of two dermal substitutes—a crosslinked and a non-crosslinked collagen biomatrix—applied topically on full-thickness skin defects paravertebrally on the back of female Göttingen Minipigs. Although positive results demonstrate that the single biomatrix application might be used in a clinical routine with small wounds, the authors conclude that the overarching aim is still the development of an innovative skin substitute to manage surface reconstruction without additional skin grafting.

Belgio et al. [2] provide an overview of retinal anatomy and diseases and a comprehensive review of retinal regeneration approaches. Scaffold-free approaches such as gene therapy and cell sheet technology are presented, as well as the fabrication techniques that can be used to produce retinal scaffolds with a particular emphasis on recent trends and advances in fabrication techniques, such as electrospinning, 3D bioprinting and lithography for retinal regeneration.

Citation: Boschetti, F. Tissue Mechanics and Tissue Engineering. *Appl. Sci.* **2022**, *12*, 6664. https://doi.org/10.3390/app12136664

Received: 24 June 2022
Accepted: 27 June 2022
Published: 30 June 2022

Nurhayati et al. [3] propose a novel design of a three-dimensional (3D) scaffold for encapsulating CD34+ cells and the idea of co-culturing CD34+ cells from different donors. Microencapsulated CD34+ cells showed no toxicity for surrounding CD34+ cells and improved their proliferation. The microencapsulation of non- or low-matched CD34+ cells is therefore proposed for protection against immune rejection and to facilitate paracrine excretion.

Forneris et al. [4] extensively investigated the structural and biomechanical heterogeneity of human abdominal aortic aneurysm tissue along the length and circumference of the aorta by means of regional ex vivo and in vivo properties. Results uniquely show the importance of regional characterization for aortic assessment and the need to correlate heterogeneity at the tissue level with non-invasive measurements aimed at improving clinical outcomes.

Messenio et al. [5] compared the intraocular pressure (IOP) measured by the Goldmann applanation tonometer (GAT) and by a pressure transducer inserted into the anterior chamber of pig eyes in vitro. Mechanical properties were also measured for the corneas of the analyzed pig eyes. As a novel result, statistical analysis revealed a correlation between the corneal mechanical properties and IOP measured by GIOP. Obtained data showed a discrepancy between the values of IOP measured by GAT and by the pressure transducer, more evident for softer and thinner corneas, which is very important for glaucoma detection.

Funding: This research received no external funding.

Acknowledgments: I would like to thank all the authors that accepted the invitation, carefully prepared and submitted their articles, and contributed to the significant advancement of knowledge in the field of TE and TM. I also acknowledge the reviewers who dedicated time and expertise to improve the quality and scientific soundness of the manuscripts. Finally, I am very grateful to the *Applied Sciences* Editorial Board for the opportunity to propose and publish this Special Issue and, in particular, to Section Managing Editor for her patience, continuous support, and assistance.

Conflicts of Interest: The author declares no conflict of interest.

References

1. Eisler, W.; Baur, J.; Held, M.; Rahmanian-Schwarz, A.; Daigeler, A.; Denzinger, M. Assessment of Two Commonly used Dermal Regeneration Templates in a Swine Model without Skin Grafting. *Appl. Sci.* **2022**, *12*, 3205. [CrossRef]
2. Belgio, B.; Salvetti, A.; Mantero, S.; Boschetti, F. The Evolution of Fabrication Methods in Human Retina Regeneration. *Appl. Sci.* **2021**, *11*, 4102. [CrossRef]
3. Nurhayati, R.; Cahyo, R.; Pratama, G.; Anggraini, D.; Mubarok, W.; Kobayashi, M.; Antarianto, R. Alginate-Chitosan Microencapsulated Cells for Improving CD34+ Progenitor Maintenance and Expansion. *Appl. Sci.* **2021**, *11*, 7887. [CrossRef]
4. Forneris, A.; Nightingale, M.; Ismaguilova, A.; Sigaeva, T.; Neave, L.; Bromley, A.; Moore, R.; Di Martino, E. Heterogeneity of Ex Vivo and In Vivo Properties along the Length of the Abdominal Aortic Aneurysm. *Appl. Sci.* **2021**, *11*, 3485. [CrossRef]
5. Messenio, D.; Ferroni, M.; Boschetti, F. Goldmann Tonometry and Corneal Biomechanics. *Appl. Sci.* **2021**, *11*, 4025. [CrossRef]

Article

Heterogeneity of Ex Vivo and In Vivo Properties along the Length of the Abdominal Aortic Aneurysm

Arianna Forneris [1,2], Miriam Nightingale [1,3], Alina Ismaguilova [1,3], Taisiya Sigaeva [4], Louise Neave [1,3], Amy Bromley [5], Randy D. Moore [6] and Elena S. Di Martino [1,2,3,*]

1 Biomedical Engineering Program, University of Calgary, Calgary, AB T2N 1N4, Canada; arianna.forneris@ucalgary.ca (A.F.); miriam.nightingale@ucalgary.ca (M.N.); aismagui@ucalgary.ca (A.I.); louise.neave@ucalgary.ca (L.N.)
2 Department of Civil Engineering, University of Calgary, Calgary, AB T2N 1N4, Canada
3 Libin Cardiovascular Institute of Alberta, University of Calgary, Calgary, AB T2N 1N4, Canada
4 Department of Systems Design Engineering, University of Waterloo, Waterloo, ON N2L 3G1, Canada; tsigaeva@uwaterloo.ca
5 Department of Pathology and Laboratory Medicine, University of Calgary, Calgary, AB T2N 1N4, Canada; Amy.Bromley@albertaprecisionlabs.ca
6 Department of Surgery, University of Calgary, Calgary, AB T2N 1N4, Canada; rmoor@ucalgary.ca
* Correspondence: edimarti@ucalgary.ca

Abstract: The current clinical guidelines for the management of aortic abdominal aneurysms (AAAs) overlook the structural and mechanical heterogeneity of the aortic tissue and its role in the regional weakening that drives disease progression. This study is a comprehensive investigation of the structural and biomechanical heterogeneity of AAA tissue along the length and circumference of the aorta, by means of regional ex vivo and in vivo properties. Biaxial testing and histological analysis were performed on ex vivo human aortic specimens systematically collected during open repair surgery. Wall-shear stress and three-dimensional principal strain analysis were performed to allow for in vivo regional characterization of individual aortas. A marked effect of position along the aortic length was observed in both ex vivo and in vivo properties, with the central regions corresponding to the aneurysmal sac being significantly different from the adjacent regions. The heterogeneity along the circumference of the aorta was reflected in the ex vivo biaxial response at low strains and histological properties. Present findings uniquely show the importance of regional characterization for aortic assessment and the need to correlate heterogeneity at the tissue level with non-invasive measurements aimed at improving clinical outcomes.

Keywords: abdominal aortic aneurysm; biaxial testing; mechanical properties; in vivo strain; wall shear stress; inflammation; regional variations; heterogeneity

Citation: Forneris, A.; Nightingale, M.; Ismaguilova, A.; Sigaeva, T.; Neave, L.; Bromley, A.; Moore, R.D.; Di Martino, E.S. Heterogeneity of Ex Vivo and In Vivo Properties along the Length of the Abdominal Aortic Aneurysm. *Appl. Sci.* **2021**, *11*, 3485. https://doi.org/10.3390/app11083485

Academic Editor: Federica Boschetti

Received: 28 February 2021
Accepted: 7 April 2021
Published: 13 April 2021

1. Introduction

An abdominal aortic aneurysm (AAA) is a slowly progressing disease that affects the wall of the abdominal aorta and results in the asymptomatic enlargement of the artery until rupture, which is associated with high mortality. The clinical management of aortic aneurysms is challenging and mostly based on an assessment aimed at weighing the risk of surgery-related complication versus the risk of catastrophic rupture. The maximum aortic diameter (greater than 5 cm) is considered the indicator for elective aortic repair; however, this approach has been proved to lead to suboptimal patient prioritization resulting in many cases of critical aneurysms left untreated or, in contrast, unnecessary interventions on stable aneurysms.

In recent years, different studies highlighted the inadequacy of the diameter criterion by pointing out the high level of heterogeneity in the aneurysmal tissue and its impact on the risk of rupture of individual aortas [1–5]. Aneurysm initiation and progression are multifactorial processes linked to the tissue's heterogeneous remodeling and structural degradation in response to the local environment, including both mechanical (altered hemodynamics) and biological factors (inflammation, presence of intraluminal thrombus) [6–8]. Considering only the diameter as a metric for rupture risk fails to capture the localized structural weakening and decrease in wall strength that drive aneurysm growth and rupture in individual aortas. In this context, the aortic diameter provides clinicians with limited information and ultimately does not account for inter- and intra-patient heterogeneity. It is therefore essential to fully characterize the local structural and mechanical changes in the aneurysmal tissue with respect to disease progression and rupture potential. This will assist in improving the aortic assessment for clinical purposes through correlation of the heterogeneity at the tissue level with non-invasive measurements.

Uniaxial tensile tests have been used to study the mechanical behaviour of aortic tissue ex vivo [9–11]. However, planar biaxial tests better represent the in vivo loading conditions of the artery and allow for a more appropriate characterization of the three-dimensional mechanical response of the aorta given tissue anisotropy and coupling of fibers in two orthogonal directions. Due to the specifics of AAA-related interventions resulting in minimal tissue being excised, as well as the overall AAA tissue fragility, there is a limited number of biaxial studies in the area [12–15]. Even less studies are concerned with heterogeneity of the AAA wall, with the focus being towards the heterogeneity along the circumference of the aorta [16]. Axial heterogeneity (along the length of the aorta) of ex vivo mechanical properties is absent in the literature. Therefore, a comprehensive assessment of the heterogeneity in both circumferential and axial directions is necessary to fully understand the localized changes in the aneurysmal wall.

As the mechanical behavior of the aorta depends heavily on its microstructure, the assessment of microstructural components in the aortic wall—in terms of content as well as architecture and organization—is essential to fully understand the pathological remodeling associated to disease progression and material properties change. While a reported increase in the collagen-to-elastin ratio is thought to alter the wall structure and thus alter its mechanical response, there is no literature data investigating the heterogeneity of the aortic wall composition along the length and circumference of the aorta [2,17,18].

Immunohistochemical (IHC) analysis provides important information on cellular mechanisms of AAA development and progression. Inflammation and macrophage infiltration are thought to have a significant role in the progressive degradation and remodeling of the extracellular matrix [19]. The primary cells involved in AAA inflammatory response are T-cells and macrophages [20]. Researchers have correlated the distribution of these cells with aneurysm rupture site, intra-luminal thrombus development, and overall disease severity [21]. IHC analysis has been used previously to investigate the regional distribution of inflammatory cells in AAA tissue [16]. However, only circumferential heterogeneity has been reported.

The invasive assessment of the aortic tissue, although central in improving the understanding of disease progression and aortic rupture, represents the first step towards the improvement of clinical guidelines for an accurate, patient-specific evaluation of rupture risk. Equally important is the need to obtain non-invasive means to access information on the state of individual aortas. To this effect, there have been several efforts to model the biomechanics of AAAs, both fluid and solid, that have led to so-called biomechanics-based indices as a surrogate measure to estimate growth and rupture risk [4,5,22,23]. The heterogeneity along the length and circumference of the aorta should be incorporated in these parameters.

In the current study, biaxial testing and histological analysis were performed on human aortic specimens ex vivo. Tissue samples were obtained from a population of open surgical repair patients in order to characterize the heterogeneity of mechanical properties and inflammatory processes of the aneurysmal tissue along the length and circumference of the aorta. Similarly, a series of non-invasive in vivo image-based parameters were obtained along the length and circumference of each aneurysm, namely computational fluid dynamics-based wall-shear stress and three-dimensional principal strain. Hence, this work provides a comprehensive investigation of the structural and biomechanical heterogeneity of the abdominal aortic aneurysm.

2. Materials and Methods

2.1. Patients and Aortic Tissue Samples

Aortic tissue samples were obtained from a population of AAA patients who consented to participate in the study between 2016 and 2020. The research protocol, approved by the University of Calgary Conjoint Health Research Ethics Board (Ethics ID: REB15-0777) included pre-operative electrocardiography-gated dynamic computed tomography (CT) imaging and open surgical aneurysm repair with complete aortic resection that allowed for the collection of specimens from different regions (i.e., aneurysm sac, aneurysm neck, anterior, posterior, left, right) of individual aortas. A sampling grid with 24 regions (or patches) was utilized in the systematic harvesting of aortic samples. Patches were defined perpendicularly to the lumen centerline of each aortic geometry reconstructed from the first phase of the dynamic CT scans (Figure 1a). Specifically, the 24 regions resulted from the subdivision of each aortic geometry into 6 regions along the length of the vessel ('1'—neck, '6'—bottom; average patch length along the centerline was 20 $\pm$ 4 mm) and 4 regions along the circumference ('L' and 'R'—left and right sides of the aneurysm, 'P' and 'A'—anterior and posterior sides of the aneurysm with respect to the geometry centerline). Both the lumen and outer aortic wall were reconstructed by means of image processing and semi-automatic, threshold-based image segmentation with the imaging software Simpleware ScanIP (Synopsys Inc., Mountain View, CA, USA). The use of a sampling grid was central to the study as it facilitated tracking of specimen locations and regional analysis of both ex vivo and in vivo properties.

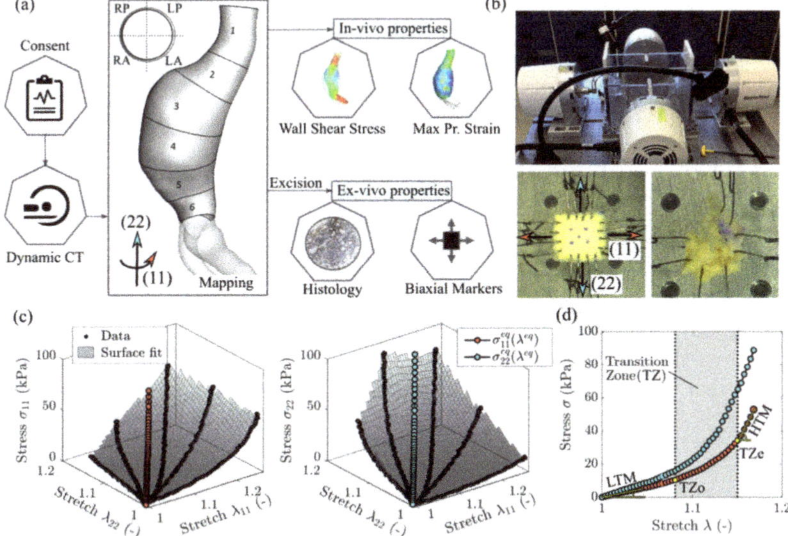

Figure 1. (**a**) Methodology summary and demonstration of the sampling grid defining the circumferential regions (RP—right posterior, LP—left posterior, RA—right anterior, LA—left anterior) and axial regions (1 to 6) of an aneurysm. (**b**) Biaxial testing setup and representative specimens: normal-sized specimen with four hooks per side and smaller specimen with two hooks per side. The latter is shown to fail due to tissue fragility. (**c**) A 3D representation of the biaxial data for a displacement-controlled test with different loading ratios. The surface interpolation is used to determine the equi-biaxial stretch-stress state. Stress and stretch are defined in the circumferential direction (11) and the longitudinal direction (22). (**d**) A 2D representation of the equi-biaxial stretch state and determination of all relevant mechanical properties—the low and high tangential modulus (LTM and HTM), the transition zone onset and end (TZo and TZe).

2.2. *Ex Vivo Analysis*

2.2.1. Biaxial Testing and Data Analysis

Biaxial testing was carried out on a four-motor biaxial testing system (ElectroForce Systems, TA Instruments, Springfield, MO, USA, Figure 1b), designed to allow independent control of each motor. Full-thickness tissue samples were cut into squares (target dimensions—10 × 10 mm) and mounted to the four linear motors via sutures and hooks (Figure 1b). Four hooks per side were used to minimize edge effects and ensure homogeneous strain distribution at the center of specimens [24]. In the case of samples with dimensions too small to accommodate four hooks, two hooks per side were used as recommended in Slazansky et al., 2016 [24]. Samples were carefully oriented with direction (11) corresponding to the direction along the circumference of the aorta (circumferential direction), and direction (22) along the length of the aorta (longitudinal direction). The thickness and side-to-side distances (where sides were assumed along the points of hooks attachment) were measured using a caliper. Five dots were drawn on the sample central region using a surgical skin marker to provide distinguishable marks that can be tracked by the overhead camera. A high-resolution digital video extensometer (DVE) camera (640 × 480 pixel resolution, 55 m focal length, 200 frames per second) mounted above the test specimen was used to track the dots, producing deformation measures at the central region of the sample. When testing a tissue's response, the in vivo environment must be mimicked. As such, the sample was fully immersed in a (PBS) solution at 37 °C and pH 7.4.

Prior to testing, a pre-load of 0.05 N was applied to avoid sagging effects due to sutures slack. Loading of the sample was recorded using two load cells (22 N). At first, due to AAA tissue fragility and a high risk of premature tearing, displacements resulting in a simultaneous 20% increase of the distances between parallel sides of the sample were assigned. Then, to achieve a wider range of mechanical responses, the sample was extended with different ratios of these initial displacements. Next, the procedure was repeated for larger loadings, namely 40% and 60%. For each test, samples were conditioned by running the test for 5 cycles at a rate of 0.6 mm/s. Over 30% of the specimens failed prematurely due to disruption of tissue by hook tear-through or delamination (Figure 1b). For the surviving specimens, the data was collected from the last cycles of the tests associated with the 40% displacement protocol.

The principal stretches λ_{11} and λ_{22} were calculated from the deformations of the center of the specimen enclosed by the tracked markers [25]. The Cauchy stresses σ_{11} and σ_{22} were determined from the forces required to displace the specimen's sides along two directions multiplied by the corresponding principal stretch and divided by the undeformed area (i.e., the corresponding side length times thickness). The three-dimensional stress–stretch curves derived from the biaxial tests for a representative specimen are shown in Figure 1c. We specify that the assigned displacements do not necessarily produce equivalent deformation data points (λ_{11}, λ_{22}) for different specimens because we do not control λ_{11}:λ_{22} ratios [25]. In order to achieve mutual comparison between specimens, one option is to apply constitutive modelling and produce comparable stress–stretch data points, for which different effective mechanical properties such as stiffness and transition strains could be determined. To bypass the constitutive modelling stage, we reached the same result by fitting a surface to the data in the three-dimensional space and interpolating the stresses σ_{11}^{eq} and σ_{22}^{eq} at the equi-stretch deformation state (i.e., $\lambda_{11} = \lambda_{22} = \lambda^{eq}$). This analysis was conducted in MATLAB (version R2018a; MathWorks, Natick, MA, USA). The equi-biaxial mechanical response for a representative dataset is shown in Figure 1 in both 3D (Figure 1c) and 2D (Figure 1d) versions. Repeating the procedure for all the specimens allows comparison of the tissues' mechanical properties at an equivalent deformation state. This approach represents an attractive alternative to constitutive modelling when finite element simulations are not the end goal.

The curves obtained from testing were then used to calculate the mechanical properties using a custom software pipeline developed in Python (version 3.7.6; Python Software Foundation). All mechanical parameters are shown in Figure 1d for an example of unprocessed equi-biaxial stress-stretch response in the circumferential direction (11). Data was first processed using a Savitsky–Golay digital filter, which relies on a local least-squares polynomial approximation to reduce noise without distorting the signal [26]. The linear regions of the curves can be described with a low-strain tangential modulus (LTM) and high-strain tangential modulus (HTM) which characterize the elastin-based and collagen-dominated response of the tissue, respectively [27,28]. These two linear regions are separated by a non-linear transition zone where the load is shared by varying proportions of collagen and elastin (Figure 1d). To determine LTM, HTM, and the transition zone, a polygonal approximation of the curve was generated using a Ramer–Douglas–Peucker (RDP) algorithm [29] with a fixed deviation value of 0.025 to produce a high-precision fit. LTM and HTM were evaluated as the slopes of the first and last segment, respectively, while the transition zone onset (TZo) and transition zone end (TZe) were found as the start and end points of the transition zone between the LTM and HTM segments (Figure 1d). For curves with two segments or less, there was deemed to be no transition zone. A visual representation of RDP and original curves were overlaid to confirm fit. Given the marked difference in AAA mechanical behavior with respect to direction, the level of mechanical anisotropy was also investigated for each mechanical parameter (MP) extracted from the biaxial response of the tissue samples as the ratio of the response in the circumferential direction to the response in the axial direction (MP_{11}/MP_{22}).

2.2.2. Histological Analysis

Histological analysis involving Musto/Movat pentachrome staining to evaluate elastin, proteoglycans, and smooth muscle cells content was performed at the Calgary Laboratory Services (Core Pathology Laboratory). Aortic specimens were fixed with 10% buffered formaldehyde solution immediately after collection, embedded in paraffin on edge and cut at 4 µm sections. Each stained sample was then analyzed by means of colorimetric analysis through the Aperio ImageScope software (version 12.3; Leica Biosystems Inc., Concord, ON, Canada). A positive pixel algorithm, validated through the Core Pathology Laboratory, was used to detect the color black, cyan, and magenta to assess the relative area of elastin, proteoglycans and vascular smooth muscle cells, respectively, and estimate their content in the specimens. The results of the colorimetric analysis was reported as a stained area (mm^2) within the total analyzed region (mm^2).

2.2.3. Immunohistochemical Analysis

IHC analysis involving immunostaining was performed at the Calgary Laboratory Services (Core Pathology Laboratory). Aortic tissue samples were fixed with 10% buffered formaldehyde solution upon harvesting in the operating room, embedded in paraffin, on edge, and cut at 4 µm sections. A Dako EnVision Flex Detection Kit (Agilent Technologies, Santa Clara, CA, USA) was used for the evaluation of inflammatory cells; specifically, the presence of the following markers was quantified: helper T-cells (CD4+), cytotoxic T-cells (CD8+), and macrophages (CD68+) [4]. This ready to use kit was used to the manufacturer's specifications (no dilution or secondary antibody retrieval necessary) with batch controls (tonsil). Analysis for CD4 (SP35) was run with H15 × 15 (15 min antibody incubation, 15 min detection kit incubation) at 1/25, for CD8 (C8/144B) with H20 × 20, and for CD68 (KP1) with H15 × 20. IHC analysis was performed in order to characterize and quantify the level of inflammation in the media and adventitia layers as the mean number of cells per 1 mm^2 specimen area. The immunostaining was counterstained with hematoxylin allowing the discrimination of the adventitia layer from the media layer. The presence of each marker was evaluated by manually counting the positively stained cells in a 1 mm^2 area identified as a hot-spot with the highest density of stained cells.

2.3. *In Vivo Analysis*

A series of in vivo analyses were performed to characterize each patch defined on the aortic geometries and allow for the correlation of in vivo and ex vivo measures on corresponding regions. For this reason, each descriptor derived from the in vivo analysis was obtained as a continuous distribution on the aorta's surface geometry as well as region-averaged distribution on the 24 regions defined on each geometry as an intra-operative sampling grid.

2.3.1. Wall Shear Stress

The evaluation of the wall shear stress was performed by means of computational fluid dynamic (CFD) simulations in order to characterize the local fluid dynamic patterns and quantify local hemodynamic disturbances. The reconstructed geometry of the aortic lumen for individual aortas was imported in Icem (version 2019.2; Ansys, Canonsburg, PA, USA) and discretized into tetrahedral elements with a final mesh density of approximately 2 to 3 million elements selected following a mesh sensitivity analysis. The use of a boundary layer consisting of prismatic elements allowed for an improved result accuracy at the geometry boundary where the variable of interest, the wall shear stress, was computed. The CFD commercial software Fluent (version 2019.2; Ansys, Canonsburg, PA, USA) was used to run transient-time simulations and reproduce the aortic fluid dynamics during a cardiac cycle. A semi-implicit method for pressure-linked equations (SIMPLE algorithm) was employed to couple pressure and velocity and solve the Navier–Stokes equations for the blood modelled as an incompressible, Newtonian fluid undergoing laminar flow [4,5]. A second order upwind formulation and a second order implicit transient formulation were used for spatial and temporal discretization respectively. Boundary conditions were imposed in order to enable a solution of the fluid flow equations. Specifically, a velocity boundary condition was defined at the computational domain inlet [30], an outflow boundary condition was prescribed at the outlets with 50% flow division into each of the iliac arteries and a no-slip condition was imposed at the wall of the fluid domain (fluid interface) assumed as rigid. The results of CFD simulations were post-processed to derive the time-averaged wall shear stress (TAWSS) representing the shear-stress loading on the aortic wall averaged over the cardiac cycle as expressed by the equation

$$\mathrm{TAWSS} = \frac{1}{\mathrm{T}} \int_0^{\mathrm{T}} |\mathrm{WSS(s,t)}|\mathrm{dt}, \tag{1}$$

where T is the duration of the cardiac cycle and WSS(s,t) is the magnitude of the wall shear stress vector at a specific location (s) and time (t).

2.3.2. Maximum Principal Strain

Three-dimensional principal strain analysis was performed in vivo on dynamic CT images by means of proprietary software ViTAA (Virtual Touch Aortic Aneurysm—patent WO-2018/068153-A1) [31,32]. The strain analysis relied on an optical flow-based algorithm and used a triangular surface mesh of the outer aortic wall as a 3D tracking model in order to measure the displacement of each mesh node over the cardiac cycle and derive the deformation gradient. Finally, the in vivo principal strain was computed from the Green-Lagrange tensor [31,32].

2.4. Statistical Analysis

The statistical analysis was completed using a custom software pipeline developed using the SciPy, Pingouin, and Statsmodels packages in Python [33–35]. The normality of all continuous variables was assessed using the Shapiro–Wilk test. Variables that did not pass an initial significance threshold of 0.05, were subjected to a Box–Cox transformation. Normality was also assessed visually through histograms and QQ (quantile-quantile) plots. The equivalence of variance for each continuous variable along position (categorical) was assessed through the Levene test with a threshold of 0.05. A linear mixed effect model was used to assess the relationships presented in this study, with inter-patient variability set as a random effect. The significance threshold was set to 0.05 for all models. Due to the robustness of these models to violations of the distributional assumptions, the significance for the Shapiro–Wilk test after the Box–Cox transformation was set to 0.1 [36]. The normality and heteroscedasticity of the residuals of all linear mixed effect models were assessed through the Shapiro–Wilk and the White test respectively.

3. Results

A total of 12 patients referred to elective AAA open repair surgery gave consent to be enrolled in the study (mean age 65 ± 6 years, 92% males). Hypertension was reported in 7 patients in the population (58%). Four patients (33%) presented dyslipidemia, with one patient also presenting coronary artery disease, while two patients (17%) in the population presented peripheral vascular disease.

The average maximum diameter for the population was 4.94 ± 0.85 cm, with 4 patients (33%) selected for surgery because they presented critical complications (e.g., penetrating atherosclerotic ulcer, acute ischemic leg, rapid aortic growth, peripheral vascular disease) despite having a maximum aortic diameter lower than 5 cm (i.e., current clinical threshold used for surgical decision). The study protocol included open repair procedure with complete aortic resection resulting in the aorta not being opened during surgery with consequent minimal blood loss for the patients. The procedure led to an average length of stay of 7.7 days for the patients, consistently with the average length of stay reported for cohorts receiving open repair surgery without complete aortic resection.

The shape of the AAA varied across patients. Position 1 was associated with the neck for all patients and exhibited no dilatation; this region was considered a non-dilated tissue control in the longitudinal heterogeneity analysis. Position 2 exhibited no dilatation in 5/12 patients with the remaining patients having some dilatation beginning in this area. Position 3, 4, and 5 were the central regions of the aneurysm and exhibited the most growth with the largest dilatation being present in position 3 for 6/12 patients, position 4 for 4/12 patients, and position 5 for 4/12 patients. In some cases where the maximum dilatation spanned multiple regions, both positions were accounted for. For position 6, 6/12 patients exhibited a decrease in dilatation while for the rest of the patients, this area remained largely dilated. A double aneurysm, where two separate dilatations are present along the abdominal aorta, was seen in 3/12 patients, with the first generally present around position 3 and another area of growth at position 5 and 6.

The results reported below concern the changes of selected ex vivo and in vivo properties along the length and circumference of the aorta. Only statistically significant results are reported and discussed.

3.1. Ex Vivo Analysis

3.1.1. Biaxial Mechanical Response of the Aortic Tissue

With approximately 30% of the specimens failing due to rupture during experimental set-up or during the initial testing protocol before reaching the 40% displacement protocol, a complete biaxial test was feasible for 24 aortic samples from 7 aneurysms. The average LTM characterizing the linear region of the tissue's response for the analysed specimens was 298.3 ± 267.3 KPa and 359.8 ± 380.7 KPa in the circumferential (11) and longitudinal (22) direction respectively, with an average anisotropic index of 1.73 ± 4.31. The average HTM was found to be 3121.0 ± 1666.1 KPa in the circumferential direction and 3381.3 ± 2605.4 KPa in the longitudinal direction, with an average anisotropic index of 2.27 ± 5.43. The tangential modulus at both low and high strain was not significantly different in the circumferential versus longitudinal direction.

The LTM did not present a significant effect of position along the aortic length, but showed a statistically significant effect of position along the circumference (Figure 2a,b). Samples collected from the left regions of the aorta exhibited higher LTM in the circumferential direction ($p = 0.039$ for left-posterior versus right-anterior; $p = 0.061$ for left-posterior versus right-posterior) and higher LTM in the longitudinal direction ($p = 0.044$ for left-anterior versus right-anterior; $p = 0.027$ for left-anterior versus right-posterior).

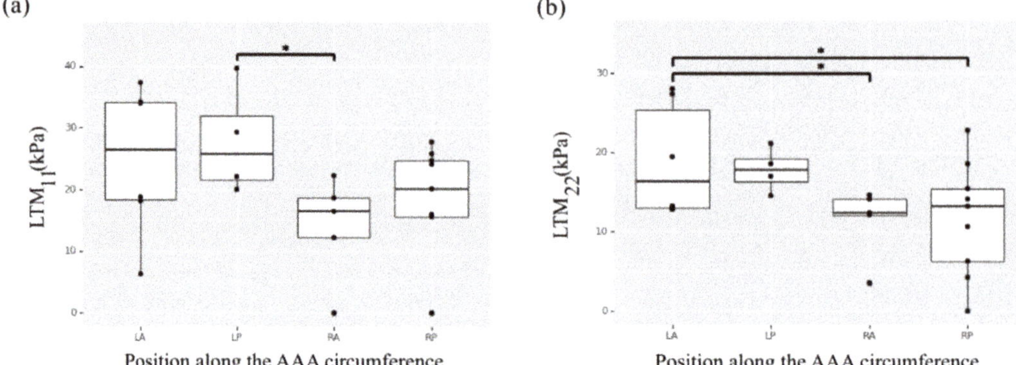

Figure 2. Heterogeneity of the selected ex vivo biaxial mechanical properties along the circumference of the AAA. (**a**) Low tangential modulus (LTM) in the circumferential direction (11). (**b**) Low tangential modulus (LTM) in the longitudinal direction (22). * indicates p value < 0.05.

Of note, a statistically significant effect of position along the aneurysm length was observed for the circumferential HTM (Figure 3a). In particular, the aneurysm neck region in position 1 exhibited higher circumferential HTM compared to the aneurysmal sac regions corresponding to position 3 (p = 0.001), position 4 (p = 0.053), and position 5 (p = 0.010). An effect of position was also observed for the HTM anisotropic index. The index was transformed through an exponential function to obtain a normal distribution. Thus, a positive number indicates a preferred fiber directionality in the circumference direction while a negative number is associated with axially oriented fibers. As these numbers approach 0, the material behaves more isotropically. The samples collected from regions in position 3 showed lower anisotropy and reached statistical significance when compared to the mechanical response characterizing position 2 or 4 (p < 0.05) (Figure 3b).

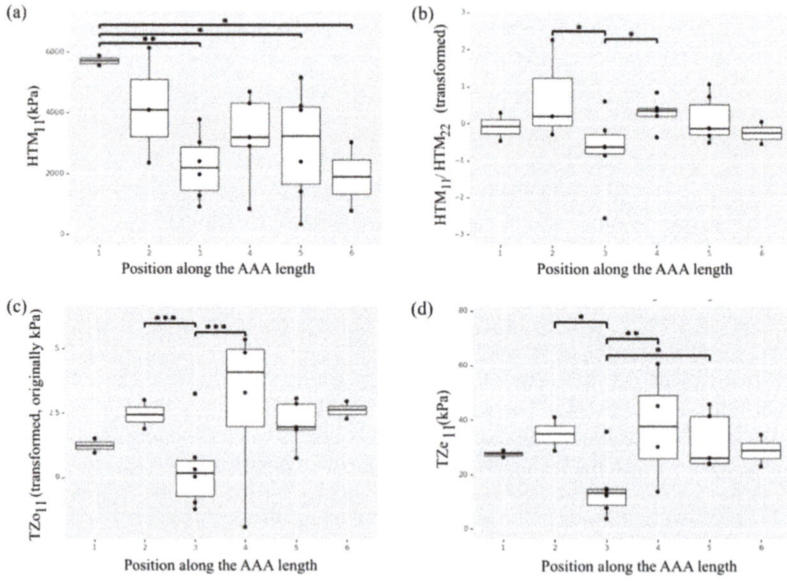

Figure 3. Heterogeneity of the selected ex vivo biaxial mechanical properties along the length of the AAA. (**a**) High tangential modulus (HTM) in the circumferential direction (11). (**b**) High tangential modulus anisotropy. (**c**) Transition zone onset (TZo) in the circumferential direction (11). (**d**) Transition zone end (TZe) in the circumferential direction (11). * indicates p value < 0.05, ** indicates p value < 0.01 and *** indicates p value < 0.001.

The non-linear transition zone of the stress-stretch curve for each AAA sample was described by means of TZo and TZe. The average TZo for the analyzed specimens was 5.3 ± 5.1 KPa in the circumferential direction and 6.1 ± 4.6 KPa in the longitudinal direction, with an average anisotropic index of 2.31 ± 6.36. The average TZe was found to be 27.8 ± 13.8 KPa and 35.0 ± 28.4 KPa respectively in the circumferential and longitudinal direction (average anisotropic index was 1.52 ± 1.84). The effect of position along the length of the artery on the parameters describing the transition zone was observed for the circumferential direction (Figure 3c,d), with samples in the central region in position 3 showing lower TZo ($p < 0.001$ compared to position 2 and 4; $p = 0.06$ compared to position 6) and TZe ($p < 0.05$ compared to position 2 and 4; $p = 0.083$ compared to position 6). The parameters describing the transition zone showed no significant effect of position along the aortic circumference.

3.1.2. Aortic Media Composition

Histological samples were collected from a subgroup of 9 patients and aortic composition was evaluated by means of Musto/Movat pentachrome staining to identify elastin, proteoglycans, and smooth muscle cells content for a total of 109 samples. Fifty-six samples (51%) had no discernible media layer and had to be excluded. Figure 4 shows a representative histology for two AAA specimens. The average elastin content for the histological specimens was 23.6% ± 10.3%, average smooth muscle cells content was 55.0% ± 16.5% and average proteoglycans content was 21.5% ± 6.5%.

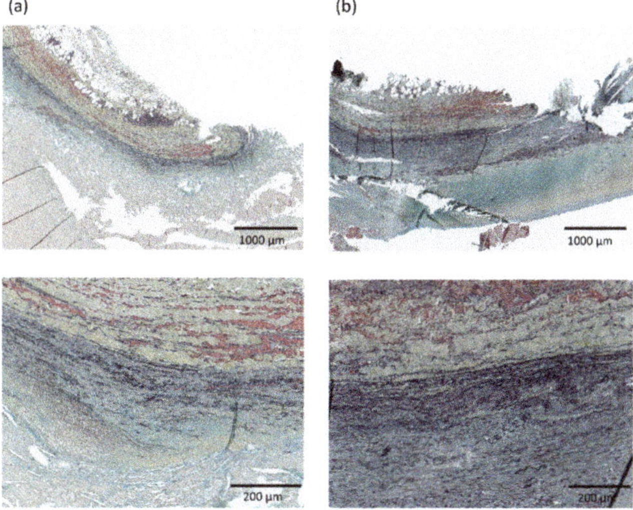

Figure 4. Representative histology showing aortic composition visualized with Musto/Movat pentachrome stain in the neck—region RA1 (column **a**) and aneurysmal dilatation—region LA3 (column **b**) for one AAA patient in the population. The layers in each histology from top to bottom are: adventitia, media, and intima. Images at 2× (top) and 10× (bottom) magnification are shown.

The composition of the media layer showed no statistically significant effect of specimen position along the length of the aorta (positions 1 to 6). A significant effect of circumferential position (right-posterior, left-posterior, right-anterior, left-anterior) on elastin, smooth muscle cells and proteoglycans content was observed (Figure 5), with specimens in anterior positions showing lower elastin content, higher smooth muscle cells content and lower proteoglycans content compared to posterior regions. Specifically left-anterior versus left-posterior, left-anterior versus right-posterior, right-anterior versus right-posterior, and right-anterior versus left-posterior ($p < 0.01$ for elastin and smooth muscle cells contents and $p < 0.04$ for proteoglycans content) (Figure 5).

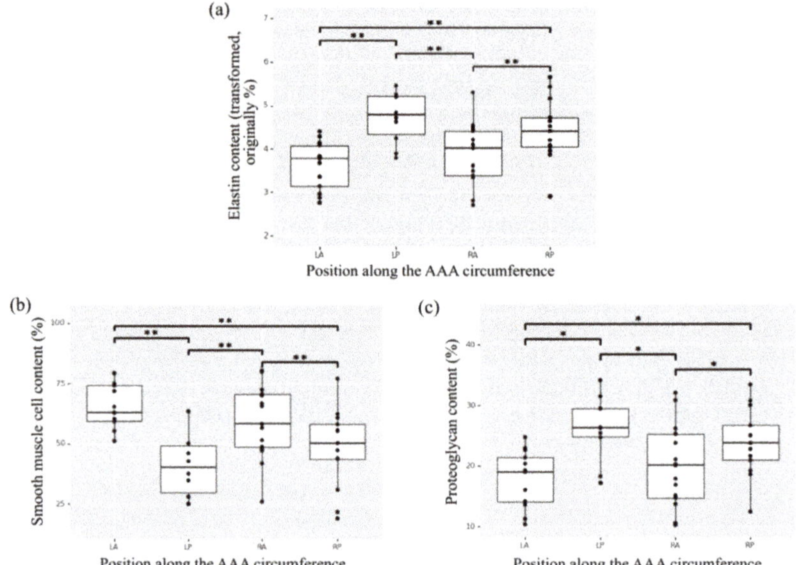

Figure 5. Heterogeneity of content along the circumference of the AAA for the aortic wall constituents (**a**) elastin, (**b**) smooth muscle cells, and (**c**) proteoglycans in the media of histological aortic samples. * indicates p value < 0.05, ** indicates p value < 0.01.

3.1.3. Markers of Inflammation

A subgroup of 7 patients allowed for the collection of aortic tissue specimens for which immunohistochemical analysis was feasible. Cell count for inflammation was therefore performed on 46 aortic samples, while several specimens had a very thin or absent media layer determining their exclusion from the analysis. The average cell count was 37. 8 ± 49.4 cells/mm^2 and 183.1 ± 429.5 cells/mm^2 for the helper T-cells (CD4+) in the media and adventitia layer respectively, 42.0 ± 49.9 cells/mm^2 and 92.8 ± 127.3 cells/mm^2 for cytotoxic T-cells (CD8+), and 22.2 ± 29.1 cells/mm^2 and 45.3 ± 37.9 cells/mm^2 for the macrophages (CD68+). Figure 6 shows a representative immunostaining image from IHC analysis with inflammatory infiltrate of CD4+, CD8+, and CD68+ cells for two AAA specimens.

Overall, the adventitia layer showed a significantly higher presence of inflammation markers compared to the media layer ($p < 0.05$ for CD4+, $p < 0.05$ for CD8+, $p < 0.01$ for CD68+). The adventitia layer also showed an effect of position observed for the CD4+ and CD8+ markers, with specimens collected from the central regions of the aorta, located in the aneurysmal dilatation, showing higher content of the two inflammatory infiltrates (Figure 7a,b). Of note, the samples collected from region 2 corresponding to the aneurysm neck showed statistically significant lower presence of CD4+ and CD8+ markers in the adventitia layer compared to samples in position 4 ($p = 0.036$ and $p = 0.004$ for CD4+ and CD8+ respectively). No significant differences were observed in the media layer in terms of inflammation and inflammatory infiltrate content with respect to position.

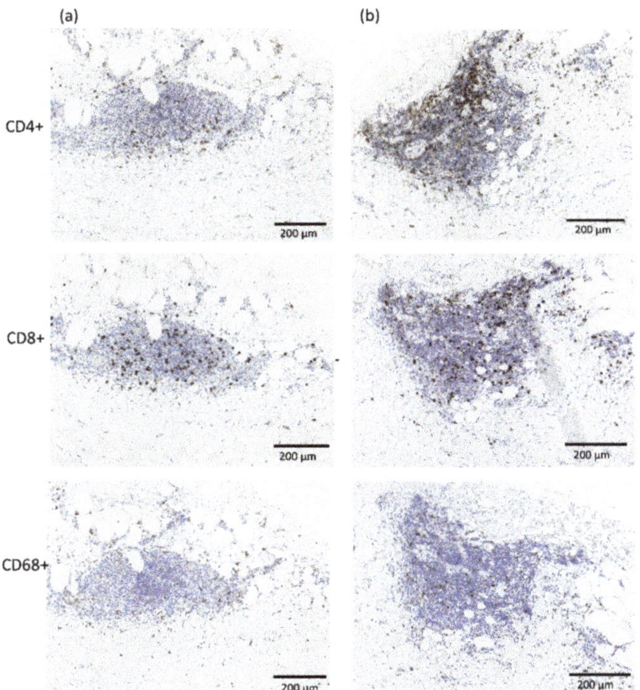

Figure 6. Representative images from immunostaining showing inflammatory infiltrate of CD4+, CD8+ and CD68+ cells in the neck—region RA1 (column **a**) and aneurysmal dilatation—region RA3 (column **b**) for one AAA patient in the population. The adventitia and media layers are in the top and bottom half of each image, respectively. All images are at 10× magnification.

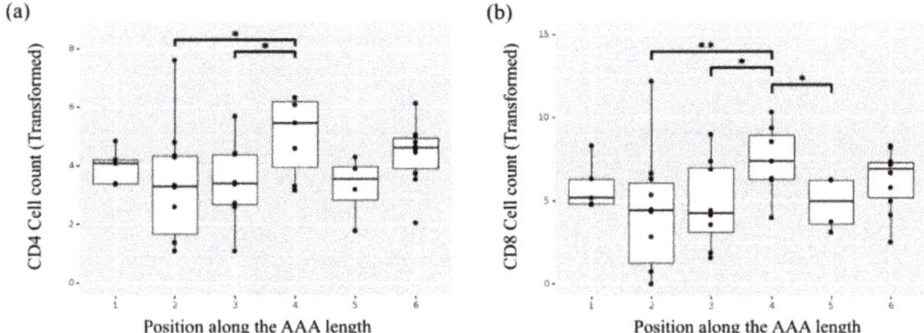

Figure 7. Heterogeneity of cell counts along the length of the AAA for the inflammatory infiltrates (**a**) helper T-cells (CD4+) and (**b**) cytotoxic T-cells (CD8+) in the adventitia layer of the selected samples undergoing ex vivo immunohistochemical analysis. * indicates p value < 0.05, ** indicates p value < 0.01.

3.2. In Vivo Analysis

The in vivo analysis was performed to obtain the CFD-based TAWSS distribution on the luminal surface of the aortic geometry and the maximum principal strain distribution derived from dynamic CT images on the outer wall surface of the artery. The in vivo analysis of maximum principal strain and wall shear stresses did not require invasive measurements and was conducted for a larger cohort of aneurysms (n = 12) including the ones selected for the ex vivo analysis. For TAWSS and maximum principal strain, a region-averaged distribution was obtained in order to characterize the same regions (24 on each aorta) with an average value.

The mean region-averaged TAWSS for the study cohort was 0.50 ± 0.24 Pa. A significant effect of position along the length of the aorta for the TAWSS was observed (Figure 8a). The central regions across the aortic geometries—namely, regions 3, 4, 5—presented lower TAWSS resulting from the slow and highly recirculating flow patterns that characterized the aneurysmal dilatation, while regions upstream (aneurysm neck region 1 and 2) and downstream the dilatation (region 6) presented a more organized, high-velocity flow that corresponded to higher TAWSS values. Of note, region 1 was characterized by statistically significant higher TAWSS compared to any other region ($p < 0.001$) while region 2 was characterized by significantly higher TAWSS compared to the more central regions in position 3 ($p = 0.004$) and position 4 and 5 ($p < 0.001$). The same central regions showed significantly lower TAWSS with respect to the areas in position 6 downstream the aneurysm sac ($p = 0.023$ with respect to position 3, $p < 0.001$ with respect to position 4 and 5).

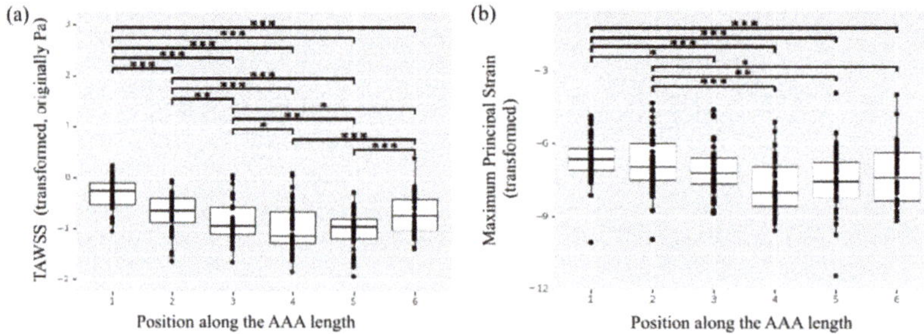

Figure 8. Heterogeneity of the in vivo properties along the length of the AAA. (**a**) TAWSS. (**b**) Maximum principal strain. * indicates p value < 0.05, ** indicates p value < 0.01 and *** indicates p value < 0.001.

The in vivo strain analysis performed on dynamic ECG-gated CT images was feasible for 11 patients in the study population, as the scan for one patient did not meet the inclusion criteria due to poor image quality. The mean region-averaged maximum principal strain for the population was 0.03 ± 0.01. Once again, the regional analysis pointed to a significant effect of position along the length of the aorta (Figure 8b), with central patches exhibiting smaller strain compared to the aneurysm neck and downstream locations. In particular, the neck region in position 1 presented a significantly larger strain compared to any of the central regions ($p = 0.02$ for region 3, $p < 0.001$ for regions 4 and 5).

Both in vivo variables of interest showed no significant effect of position along the circumference of the aorta.

An interesting relationship was found when comparing the TAWSS for the aortic regions with collected specimens showing no discernible media layer versus discernible media layer. The aortic tissue showing highly compromised composition with no clearly discernible media layer presented statistically significant lower TAWSS ($p < 0.001$) (Figure 9).

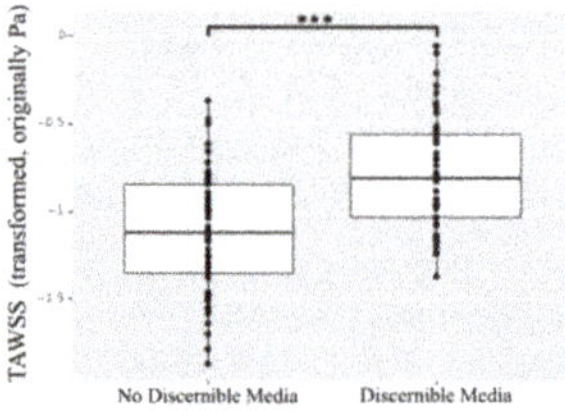

Figure 9. TAWSS for specimens showing no discernible media layer versus discernible media layer. *** indicates p value < 0.001.

4. Discussion

The present study aimed at investigating the structural and biomechanical heterogeneity of aortic aneurysmal tissue, with respect to the position along the length and circumference of the artery. Both invasive and non-invasive image-based measures were performed on regions of AAAs for a population of patients referred to elective surgery. Specifically, open repair surgery involving complete aortic resection allowed for the collection of tissue specimens from different regions along the aorta, enabling the ex vivo regional characterization of the aneurysmal tissue by means of biaxial testing, with novel biaxial parameters and histological analysis. It is important to note that studies on the ex vivo characterization of aneurysmal tissue often focus on one specimen as representative of the aneurysm or on few samples mostly collected from the anterior portion of the aorta [12]. For this reason, research dealing with regional differences has often been carried out on animal models in order to compensate for the limited availability of human tissue [14,15].

Biaxial mechanical testing of AAA tissue enables accurate assessment of the coupling of fibers in two orthogonal directions—and thus of anisotropy—spanning a range of physiological stresses/strains. However, studies focusing on biaxial testing are scarce due to aneurysmal tissue fragility. In our study, more than 30% specimens failed before reaching the 40% displacement protocol. The remaining specimens required extremely delicate handling.

Alterations in the biomechanical properties of the AAA are attributed to the changes in the structure of the extracellular matrix (ECM) constituents, mainly the elastin and collagen fibers. The low-strain tangential modulus (LTM) is associated with the behavior at low strain often attributed to the elastin component. The onset stress at the beginning of the transition zone (TZo) marks the instant when the collagen begins to engage and resist the applied loading. The end of the transition zone (TZe) is indicative of the collagen in the tissue being fully activated and the transition from the elastin-dominated mechanical response to the collagen-dominated response is complete. Finally, the high-strain tangential modulus (HTM) captures the fully engaged collagen-associated behavior.

The LTM was the only mechanical parameter found to be significantly affected by the position along the circumference of the AAA. The LTM is associated with the elastin-dominated region of aortic tissue behavior. Thus, this asymmetry could be associated with varying distributions of elastin content. In fact, the left patches of AAAs were found to have higher elastin content, which was clearly reflected in the biaxial data showing higher LTM in the left-anterior (LA) and left-posterior (LP) regions. Similarly, histological samples collected from posterior regions showed higher elastin content reflected in the trend towards higher LTM for posterior biaxial specimens. Interestingly, significant differences in LTM were associated with opposing circumferential regions.

Other studies also reported the heterogeneity along the circumferential regions of the aorta [11,37]. However, their assessment was based on uniaxial biomechanical parameters, and thus, their results are true for the high-strain mechanical response only with no consideration for the coupling of fibers in two orthogonal directions. The present study, to the best of the authors' knowledge, is the first to report circumferential heterogeneity at the low strain regime (LTM).

Regarding the axial heterogeneity, there was no effect of the position along the length of the aorta on LTM values, potentially due to the higher variability in LTM, especially in the aneurysmal sac (position 3 and 4).

The present results show that the collagen-related biaxial properties TZo, TZe, and HTM in the circumferential direction were significantly different for tissue samples collected from different regions along the length of the AAA (positions 1 to 6). This key result suggests that the pathology has a variable effect on the collagen fibers—which are oriented predominantly in the circumferential direction—along the different segments of the aneurysm resulting in longitudinal heterogeneity. Literature findings show that aneurysm progression results in the alterations of various biaxial properties along the circumferential direction (11) of AAAs [12]. However, variation of these circumferential changes along the length of AAA, to the best of the authors' knowledge, have not been addressed.

Despite the lack of studies, the longitudinal heterogeneity of the circumferential response of the tissue may be due to the large fluctuations in diameter along the aortic geometry in the presence of an aneurysmal dilatation. Collagen has a relatively short half-life and newly deposited collagen at locations corresponding to large circumferential dilatation may present lower fiber ondulation [38]. Future studies should explore how collagen fibers change along the length of the AAA.

When compared to other regions along the length of the aorta, the aneurysm neck (position 1) was found to be the stiffest, with significantly higher circumferential HTM (direction 11). The aneurysm neck's tissue could be considered as representative of the non-dilated tissue given its position upstream the enlargement. In this case, we would obtain opposite result with respect to other studies that report the aneurysmal tissue as stiffer compared to a healthy control [12]. In contrast, our results suggest that the presence of an aneurysm leads to a decrease in HTM/stiffness along the length of the aorta as it dilates within the same patient. In fact, position 3 (located in the aneurysmal sac) exhibited both the lowest HTM and the largest diameter within our investigated subset of patients. Previous studies have shown that the collagen present in the aneurysmal tissue is more disorganized, as well as thinner and more elongated, potentially resulting in lower circumferential HTM at sites of advanced disease progression versus non-dilated tissue such as the neck [39].

In contrast to previous studies [12,40], we did not observe a pronounced anisotropy when analyzing the specimens altogether. In particular, the neck (position 1) and the region downstream of the aneurysm (position 6) were observed to be isotropic. However, a significant difference was observed in the anisotropy associated with HTM with respect to the longitudinal position within the body. Position 3 had a tendency towards a preferred longitudinal fiber directionality, while the adjacent patches (position 2 and 4) exhibited preferred fiber directionality in the circumferential direction. The drastic change in anisotropy associated with position 3 indicates the disorganization of circumferentially aligned collagen fibers in the aneurysmal sac resulting in a lower circumferential HTM as seen in this study, but little change in longitudinal HTM. The heterogeneity of the tissue's mechanical response may explain the inconsistency between the present results and other studies, in which only one sample was excised from the aneurysm and, thus, was representative of only one region.

The infamous position 3 (in the aneurysmal dilatation) had the lowest TZo and TZe along the length of the artery and was significantly different compared to adjacent patches. This suggests that collagen is becoming activated and the tissue is transitioning into a collagen-dominated behavior at lower stresses. The earlier collagen activation may be due to the elastin being depleted in the area of the aneurysm compared to adjacent regions as the collagen replaces it to provide mechanical support. This transition appears to be related to the position/geometry of the aneurysm. As previously noted, position 3 generally was the most dilated area of the AAA. The outliers for TZo and TZe in position 3 and 4 originated from the same patient which presented with position 4 associated with the largest diameter.

Histological samples allowed for the analysis of aortic wall structure and composition as it relates to the mechanical properties of the aorta. More than 50% of these specimens presented non-discernible media, indicative of high degeneration and loss of structural integrity in the aortic tissue under investigation. Elastin, smooth muscle cells, and proteoglycans contents were assessed in the remaining samples. While elastin has been reported to be reduced in aneurysms [17], with ruptured AAAs presenting lower elastin content than non-ruptured ones [41], the regional heterogeneity of elastin content has not been previously reported. Present results showed no effect of the position along the length of the aorta on the content of medial constituents, but a significant effect of the circumferential position was found. Elastin and proteoglycans were observed to be present in inverse proportions with smooth muscle (i.e., an increase in relative elastin and proteoglycans is associated with a decrease in smooth muscle). Of note, specimens collected from the anterior regions of the aorta presented significantly lower elastin content compared to posterior regions. As previously discussed, the only other parameter significantly affected by the position along the circumference was the LTM that characterizes the elastin-dominated mechanical behavior of the tissue at low strain. Greater elastin content and higher LTM in posterior regions of the aorta indicate an asymmetric remodeling and degenerative process at the wall, with asymmetric growth and thrombus accumulation occurring predominantly in the anterior bulging.

Literature findings suggest that the AAA pathophysiology is a multifactorial process, with cellular mechanisms integrally involved in the structural and functional changes that occur in the aortic wall. However, relatively few studies have investigated the heterogeneity of inflammatory response in human AAA and its possible implications.

An effect of circumferential position was previously reported by Hurks et al., who found that lateral regions of AAA exhibited increased inflammatory activity in the adventitia compared to anterior and posterior regions [16]. In contrast, the present study showed no significant variation in inflammatory cell count among circumferential regions, although it should be noted that different inflammation markers were analyzed. However, when the effect of axial position was investigated, tissue samples showed significant heterogeneity in terms of inflammatory infiltrates. Higher inflammatory cell activity (CD4+, CD8+) in the adventitia was found in samples collected from regions in position 4 compared to position 2 and 3. The localized increase in helper T-cells (CD4+) and cytotoxic T-cells (CD8+) together in this region suggests the presence of CD4+ T-cell phenotype Th1, which has been implicated in ECM degradation [42]. CD4+ T-cells were also reported to release cytokines that stimulate angiogenesis and fibroblastic collagen accumulation in the adventitia. Therefore, their appearance is highly indicative of aortic wall remodeling [43,44]. Interestingly, position 4 exhibited distinct mechanical behaviour compared to position 2 and 3, with a significantly higher TZo and TZe, possibly a result of increased fibrotic collagen deposition in the adventitia in combination with a degraded media.

These results are significant in the overall discussion on AAA risk assessment: the axial heterogeneity in inflammatory cells seems to have direct functional implications that can be seen in the biomechanical results of this study.

While the ex vivo mechanical and histological characterization of the aorta is essential for a thorough understanding of the structural and functional changes linked to AAA progression, the in vivo assessment becomes central when clinical application and disease management are the goal.

From in vivo, image-based regional assessment of the AAA population, the TAWSS was found to be affected by the position along the axis of the aortic geometry with areas of altered, recirculating flow (low TAWSS) marking the central aneurysm regions as a consequence of dilatation. The shear stress resulting from the viscous nature of blood flow is unlikely to load the aortic tissue to the point of failure; however, the wall-shear stress is involved in the processes of mechano-sensing and mechano-transduction responsible for the local pathological remodeling and structural degeneration of the aortic tissue. Because of the tissue's response to site-specific hemodynamic conditions, the aortic wall presents highly heterogeneous material properties, especially in the presence of degenerative processes. The effect of wall-shear stress has been linked to disease progression before, with literature reporting on its role in thrombus formation and accumulation—likely leading to hypoxia and further loss in tissue integrity, as well as aneurysm rupture [2,4–7]. The present findings, in agreement with previous literature, highlight the aneurysm as a region of significantly disturbed flow where further degeneration of the tissue is likely to occur. As the aorta dilates, its hemodynamics is subject to additional disturbance driven by geometrical changes, making the dilatation itself a region more and more prone to local pathological processes. Interestingly, the TAWSS was found to be significantly lower for the regions characterized by tissue specimens showing no discernible media (more than

50% of the histological samples), further highlighting the relationship between local fluid dynamic patterns and tissue degeneration.

On the one hand, the study of the local hemodynamics provides an understanding of the biological substrate for the aortic wall that leads to pathological remodelling and enlargement. On the other hand, the local deformability of the wall can help further characterize the state of regional weakening of the aortic wall and its propensity for rapid dilatation or rupture. The in vivo three-dimensional strain analysis allowed for the non-invasive assessment of localized wall behavior directly from dynamic CT imageswithout assumptions made on constitutive models. The region-averaged distribution of maximum principal strain for the AAA population further supports the concept of heterogeneity in aneurysmal tissue as a result of the heterogeneous remodeling, and the significant effect of the position along the length of the aorta on its biomechanics. The central regions corresponding to the aneurysmal sac (position 3, 4, 5) presented significantly smaller strain compared to the more proximal (neck) regions.

As the in vivo local deformability of the aorta relates to local mechanical properties, a large strain may be a possible consequence of flow impingement on the aortic wall identified in regions with a corresponding high TAWSS. Similarly, in the regions of low TAWSS, such as the larger diameter areas, the deformation of the aortic wall is expected to be small, especially in presence of a thick thrombus buffering the aorta. Therefore, in these regions, a larger strain may be indicative of intrinsic localized weakening of the tissue, as the low blood velocities are unlikely to be causing the deformation of the tissue. Of note, the distribution of strain presented more variability than the distribution of wall shear stress further highlighting the dual influence of loading (flow impingement) and material weakening on the localized wall strain.

This is a pioneering work on the regional heterogeneity of the AAA that comprises both in vivo and ex vivo analyses. The comprehensive investigation highlighted novel results on the heterogeneity of the aortic tissue along both the circumference and length of the aneurysm. The heterogeneity along the circumference of the aorta was only reflected in the ex vivo biaxial response at low strains, which was linked to variations of elastin content. A particularly marked effect of position along the aortic length, in contrast, was consistently observed in several ex vivo and in vivo properties, with the central regions corresponding to the aneurysmal dilatation (particularly position 3) being strikingly distinct from the adjacent patches (i.e., showing significant difference with respect to the other identified regions). The aneurysm area was characterized by disturbed hemodynamics likely to drive the pathological remodeling that results in a changed mechanical response as observed in the heterogeneity of collagen-related behavior (HTM, TZo, and TZe) and inflammatory markers content along the length of the artery.

First, the consistency observed between ex vivo and in vivo properties clearly suggests that in vivo biomarkers can eventually improve aneurysm assessment and outcome prediction, with regional heterogeneity providing a direction for future work on non-invasive risk assessment. Second, while the central regions of the aorta corresponded to the position of the aneurysmal sac and, therefore, included the location of maximum diameter, the present research clearly points out the shortcomings of the use of maximum diameter. The aortic size alone cannot fully characterize the regional variability and heterogeneity of the AAA tissue, especially when considering that rupture has been reported to often occur away from the location of maximum diameter [4,7].

This study presents limitations that need to be addressed. The sample size was limited, in terms of both patients and AAA tissue specimens; present results would benefit from analyses performed on a larger cohort that can better represent the variability among different patients and aortic anatomies. A larger study cohort would possibly allow for the investigation of difference in female versus male AAA patients, as well as age-matched analysis, given the effect of sex and aging on the arterial structure and function. Similarly, access to healthy aortic tissue and imaging would provide a means for valuable considerations.

There are shortcomings in the colorimetric analysis performed on histologies as only three colors on a pentachrome stain were analyzed: darker pixels may be recognized as black, ultimately affecting the elastin content results. Moreover, due to the scarcity of specimens containing measurable collagen in the media, along with the faint nature of the Musto/Movat staining for collagen, analysis for this constituent was not feasible. Further studies should be conducted to include Picrosirus Red staining to allow for collagen visualization, and Total Collagen Assay and ELISA to allow for quantitative assessment of collagen in the tissue. Additionally, the use of multiphoton microscopy would prove useful to investigate collagen fiber morphology (i.e., fiber thickness, direction, ondulation) as it relates to the mechanical properties of the tissue. While the Musto/Movat staining provided an insight into the elastin content of the specimens, it did not allow for any inference on the state of the elastin fibers, thus additional analysis would prove useful to access information on elastin fragmentation. Similarly, a look into extracellular matrix breakdown and matrix metalloproteinase would provide information on the proteolytic process associated with the degradation of the aortic media.

The IHC analysis had limitations due to the presence of background staining that can lead to a misinterpreted and overestimated cell count. In this regard, manual cell counting can be prone to errors and may be insufficient as a stand-alone metric for inflammation as cytokines, signaling interactions and molecular mechanisms are also involved in the complex inflammatory process.

The assumption of rigid aortic wall for CFD simulations also presented limitations. Despite being non-realistic, this assumption is an acceptable simplification that allows the characterization of the main hemodynamic patterns given the unknown, and extremely heterogeneous, patient-specific material properties that would be needed for more computationally expensive fluid–structure interaction (FSI) simulations.

While the primary objective of the present study was the ex vivo and in vivo mechanical characterization of the aortic tissue and its heterogeneity, the sparse experimental data did not allow a one-to-one regional correspondence for each analysis, therefore limiting the investigation of relationships among the different parameters. Future work will look at prospective longitudinal studies to assess regional aortic growth as a result of localized weakening with the aim of correlating regional growth to non-invasive parameters that can be measured clinically.

5. Conclusions

The workflow and methodology described herein allowed for the characterization of the aortic tissue with a comprehensive location-specific analysis and provided a distinctive insight on the assessment of AAAs. The effect of position on ex vivo and in vivo properties was investigated and showed the significant heterogeneity present in AAA tissue along both the length and the circumference of the aorta, with the most striking and consistent results being found with respect to the position along the length of the artery.

Present findings highlight the heterogeneity of AAAs and the essential role of regional characterization in the context of aortic assessment for disease management purposes. More work should be done towards the implementation of novel approaches accounting for the heterogeneity of the aneurysm—through correlation of the heterogeneity at the tissue level with non-invasive measurements—in order to improve risk stratification and clinical outcomes for individual AAA patients.

Author Contributions: Conceptualization, E.S.D.M.; Data curation, A.F.; Formal analysis, A.F., M.N., T.S. and L.N.; Funding acquisition, E.S.D.M.; Investigation, A.F., A.I. and A.B.; Methodology, A.F., T.S. and E.S.D.M.; Project administration, E.S.D.M.; Resources, R.D.M. and E.S.D.M.; Supervision, E.S.D.M.; Validation, A.F.; Visualization, M.N. and T.S.; Writing—original draft, A.F., M.N., T.S. and L.N.; Writing—review and editing, A.F., M.N., A.I., T.S., L.N., A.B., R.D.M. and E.S.D.M. All authors have read and agreed to the published version of the manuscript.

Funding: This research was funded by the Werner Graupe International Fellowship in Engineering and the Libin Cardiovascular Institute of Alberta scholarship, with contributions from the Natural

Science and Engineering Research Council of Canada Discovery Grant RGPIN/04043-2014, and RGPIN/07178-2019, and the Heart and Stroke Foundation Grant in Aid G 170019141.

Institutional Review Board Statement: The study was conducted according to the guidelines of the Declaration of Helsinki, and approved by the University of Calgary Conjoint Health Research Ethics Board (CHREB) (Ethics ID: REB15-0777).

Informed Consent Statement: Informed written consent was obtained from all subjects involved in the study.

Data Availability Statement: The data presented in this study are available on request from the corresponding author.

Acknowledgments: The authors would like to acknowledge the work done by Richard Beddoes, Flavio Bellacosa Marotti, and Christi Findlay in support of this research study.

Conflicts of Interest: E.S.D.M., R.D.M., and A.F. are co-founders and shareholders of the start-up company ViTAA.

References

1. Vallabhaneni, S.R.; Gilling-Smith, G.L.; How, T.V.; Carter, S.D.; Brennan, J.A.; Harris, P.L. Heterogeneity of tensile strength and matrix metalloproteinase activity in the wall of abdominal aortic aneurysms. *J. Endovasc. Ther.* **2004**, *11*, 494–502. [CrossRef]
2. Martufi, G.; Forneris, A.; Nobakht, S.; Rinker, K.D.; Moore, R.D.; Di Martino, E.S. Case Study: Intra-Patient Heterogeneity of Aneurysmal Tissue Properties. *Front. Cardiovasc. Med.* **2018**, *5*, 82. [CrossRef] [PubMed]
3. Derwich, W.; Wittek, A.; Hegner, A.; Fritzen, C.P.; Blase, C.; Schmitz-Rixen, T. Comparison of Abdominal Aortic Aneurysm Sac and Neck Wall Motion with 4D Ultrasound Imaging. *Eur. J. Vasc. Endovasc. Surg.* **2020**, *60*, 539–547. [CrossRef] [PubMed]
4. Forneris, A.; Marotti, F.B.; Satriano, A.; Moore, R.D.; Di Martino, E.S. A novel combined fluid dynamic and strain analysis approach identified abdominal aortic aneurysm rupture. *J. Vasc. Surg. Cases Innov. Tech.* **2020**, *6*, 172–176. [CrossRef]
5. Forneris, A.; Kennard, J.; Ismaguilova, A.; Shepherd, R.D.; Studer, D.; Bromley, A.; Moore, R.D.; Rinker, K.D.; Di Martino, E.S. Linking aortic mechanical properties, gene expression and microstructure: A new perspective on regional weakening in abdominal aortic aneurysms. *Front. Cardiovasc. Med.* **2021**, *8*, 47. [CrossRef] [PubMed]
6. Zambrano, B.A.; Gharahi, H.; Lim, C.Y.; Jaberi, F.A.; Choi, J.; Lee, W.; Baek, S. Association of intraluminal thrombus, hemodynamic forces, and abdominal aortic aneurysm expansion using longitudinal CT Images. *Ann. Biomed. Eng.* **2016**, *44*, 1502–1514. [CrossRef]
7. Boyd, A.J.; Kuhn, D.C.S.; Lozowy, R.J.; Kulbisky, G.P. Low wall shear stress predominates at sites of abdominal aortic aneurysm rupture. *J. Vasc. Surg.* **2016**, *63*, 1613–1619. [CrossRef]
8. Meyrignac, O.; Bal, L.; Zadro, C.; Vavasseur, A.; Sewonu, A.; Gaudry, M.; Saint-Lebes, B.; De Masi, M.; Revel-Mouroz, P.; Sommet, A.; et al. Combining volumetric and wall shear stress analysis from CT to assess risk of abdominal aortic aneurysm progression. *Radiology* **2020**, *295*, 722–729. [CrossRef]
9. Raghavan, M.L.; Webster, M.W.; Vorp, D.A. Ex vivo biomechanical behavior of abdominal aortic aneurysm: Assessment using a new mathematical model. *Ann. Biomed. Eng.* **1996**, *24*, 573–582. [CrossRef]
10. Di Martino, E.S.; Bohra, A.; Vande Geest, J.P.; Gupta, N.; Makaroun, M.S.; Vorp, D.A. Biomechanical properties of ruptured versus electively repaired abdominal aortic aneurysm wall tissue. *J. Vasc. Surg.* **2006**, *43*, 570–576. [CrossRef]
11. Raghavan, M.L.; Kratzberg, J.; Castro de Tolosa, E.M.; Hanaoka, M.M.; Walker, P.; da Silva, E.S. Regional distribution of wall thickness and failure properties of human abdominal aortic aneurysm. *J. Biomech.* **2006**, *39*, 3010–3016. [CrossRef] [PubMed]
12. Vande Geest, J.P.; Sacks, M.S.; Vorp, D.A. The effects of aneurysm on the biaxial mechanical behavior of human abdominal aorta. *J. Biomech.* **2006**, *39*, 1324–1334. [CrossRef] [PubMed]
13. Vande Geest, J.P.; Sacks, M.S.; Vorp, D.A. Age dependency of the biaxial biomechanical behavior of human abdominal aorta. *J. Biomech. Eng.* **2004**, *126*, 815–822. [CrossRef]
14. Bersi, M.R.; Bellini, C.; Di Achille, P.; Humphrey, J.D.; Genovese, K.; Avril, S. Novel methodology for characterizing regional variations in the material properties of murine aortas. *J. Biomech. Eng.* **2016**, *138*, 071005. [CrossRef]
15. Bersi, M.R.; Bellini, C.; Humphrey, J.D.; Avril, S. Local variations in material and structural properties characterize murine thoracic aortic aneurysm mechanics. *Biomech. Model. Mechanobiol.* **2019**, *18*, 203–218. [CrossRef]
16. Hurks, R.; Pasterkamp, G.; Vink, A.; Hoefer, I.E.; Bots, M.L.; van de Pavoordt, H.D.W.M.; de Vries, J.P.P.M.; Moll, F.L. Circumferential heterogeneity in the abdominal aortic aneurysm wall composition suggests lateral sides to be more rupture prone. *J. Vasc. Surg.* **2012**, *55*, 203–209. [CrossRef] [PubMed]
17. Rizzo, R.J.; McCarthy, W.J.; Dixit, S.N.; Lilly, M.P.; Shively, V.P.; Flinn, W.R.; Yao, J.S.T. Collagen types and matrix protein content in human abdominal aortic aneurysms. *J. Vasc. Surg.* **1989**, *10*, 365–373. [CrossRef]
18. Busch, A.; Hartmann, E.; Grimm, C.; Ergün, S.; Kickuth, R.; Otto, C.; Kellersmann, R.; Lorenz, U. Heterogeneous histomorphology, yet homogeneous vascular smooth muscle cell dedifferentiation, characterize human aneurysm disease. *J. Vasc. Surg.* **2017**, *66*, 1553–1564.e6. [CrossRef] [PubMed]

19. Raffort, J.; Lareyre, F.; Clément, M.; Hassen-Khodja, R.; Chinetti, G.; Mallat, Z. Monocytes and macrophages in abdominal aortic aneurysm. *Nat. Rev. Cardiol.* **2017**, *14*, 457–471. [CrossRef]
20. Forester, N.D.; Cruickshank, S.M.; Scott, D.J.A.; Carding, S.R. Functional characterization of T cells in abdominal aortic aneurysms. *Immunology* **2005**, *115*, 262–270. [CrossRef]
21. Sagan, A.; Mikolajczyk, T.P.; Mrowiecki, W.; MacRitchie, N.; Daly, K.; Meldrum, A.; Migliarino, S.; Delles, C.; Urbanski, K.; Filip, G.; et al. T Cells are dominant population in human abdominal aortic aneurysms and their infiltration in the perivascular tissue correlates with disease severity. *Front. Immunol.* **2019**, *10*, 1979. [CrossRef]
22. Vande Geest, J.P.; Di Martino, E.S.; Bohra, A.; Makaroun, M.S.; Vorp, D.A. A biomechanics-based rupture potential index for abdominal aortic aneurysm risk assessment: Demonstrative application. *Ann. N. Y. Acad. Sci.* **2006**, *1085*, 11–21. [CrossRef]
23. Stevens, R.R.F.; Grytsan, A.; Biasetti, J.; Roy, J.; Liljeqvist, M.L.; Christian Gasser, T. Biomechanical changes during abdominal aortic aneurysm growth. *PLoS ONE* **2017**, *12*, e0187421. [CrossRef]
24. Slazansky, M.; Polzer, S.; Man, V.; Bursa, J. Analysis of accuracy of biaxial tests based on their computational simulations. *Strain* **2016**, *52*, 424–435. [CrossRef]
25. Sigaeva, T.; Polzer, S.; Vitásek, R.; Di Martino, E.S. Effect of testing conditions on the mechanical response of aortic tissues from planar biaxial experiments: Loading protocol and specimen side. *J. Mech. Behav. Biomed. Mater.* **2020**, *111*, 103882. [CrossRef]
26. Schafer, R.W. What is a savitzky-golay filter? *IEEE Signal Process Mag.* **2011**, *28*, 111–117. [CrossRef]
27. Holzapfel, G.A. SECTION 10.11—Biomechanics of Soft Tissue. In *Handbook of Materials Behavior Models*; Lemaitre, J., Ed.; Academic Press: Burlington, MA, USA, 2001; pp. 1057–1071.
28. Holzapfel, G.A. Collagen in arterial walls: Biomechanical aspects. In *Collagen. Structure and Mechanics*; Fratzl, P., Ed.; Springer: Berlin/Heidelberg, Germany, 2008; pp. 285–324.
29. Prasad, D.K. Geometric Primitive Feature Extraction—Concepts, Algorithms, and Applications. Ph.D. Thesis, Nanyang Technological University, Singapore, 2012.
30. Mills, C.J.; Gabe, I.T.; Gault, J.H.; Mason, D.T.; Ross, J.; Braunwald, E.; Shillingford, J.P. Pressure-flow relationships and vascular impedance in man. *Cardiovasc. Res.* **1970**, *4*, 405–417. [CrossRef] [PubMed]
31. Satriano, A.; Rivolo, S.; Martufi, G.; Finol, E.A.; Di Martino, E.S. In vivo strain assessment of the abdominal aortic aneurysm. *J. Biomech.* **2015**, *48*, 354–360. [CrossRef]
32. Satriano, A.; Guenther, Z.; White, J.A.; Merchant, N.; Di Martino, E.S.; Al-Qoofi, F.; Lydell, C.P.; Fine, N.M. Three-dimensional thoracic aorta principal strain analysis from routine ECG-gated computerized tomography: Feasibility in patients undergoing transcatheter aortic valve replacement. *BMC Cardiovasc. Disord.* **2018**, *18*, 76. [CrossRef]
33. Seabold, S.; Perktold, J. Statsmodels: Econometric and Statistical Modeling with Python. In Proceedings of the 9th Python in Science Conference, Austin, TX, USA, 28–30 June 2010.
34. Vallat, R. Pingouin: Statistics in Python. *J. Open Source Softw.* **2018**, *3*, 1026. [CrossRef]
35. Virtanen, P.; Gommers, R.; Oliphant, T.E.; Haberland, M.; Reddy, T.; Cournapeau, D.; Burovski, E.; Peterson, P.; Weckesser, W.; Bright, J.; et al. SciPy 1.0: Fundamental algorithms for scientific computing in Python. *Nat. Methods* **2020**, *17*, 261–272. [CrossRef] [PubMed]
36. Schielzeth, H.; Dingemanse, N.J.; Nakagawa, S.; Westneat, D.F.; Allegue, H.; Teplitsky, C.; Réale, D.; Dochtermann, N.A.; Garamszegi, L.Z.; Araya-Ajoy, Y.G. Robustness of linear mixed-effects models to violations of distributional assumptions. *Methods Ecol. Evol.* **2020**, *11*, 1141–1152. [CrossRef]
37. Thubrikar, M.J.; Labrosse, M.; Robicsek, F.; Al-Soudi, J.; Fowler, B. Mechanical properties of abdominal aortic aneurysm wall. *J. Med. Eng. Technol.* **2001**, *25*, 133–142. [PubMed]
38. Humphrey, J.D. Vascular adaptation and mechanical homeostasis at tissue, cellular, and sub-cellular levels. *Cell Biochem. Biophys.* **2008**, *50*, 53–78. [CrossRef]
39. Urabe, G.; Hoshina, K.; Shimanuki, T.; Nishimori, Y.; Miyata, T.; Deguchi, J. Structural analysis of adventitial collagen to feature aging and aneurysm formation in human aorta. *J. Vasc. Surg.* **2016**, *63*, 1341–1350. [CrossRef]
40. O'Leary, S.A.; Kavanagh, E.G.; Grace, P.A.; McGloughlin, T.M.; Doyle, B.J. The biaxial mechanical behaviour of abdominal aortic aneurysm intraluminal thrombus: Classification of morphology and the determination of layer and region specific properties. *J. Biomech.* **2014**, *47*, 1430–1437. [CrossRef]
41. Sakalihasan, N.; Heyeres, A.; Nusgens, B.V.; Limet, R.; Lapiére, C.M. Modifications of the extracellular matrix of aneurysmal abdominal aortas as a function of their size. *Eur. J. Vasc. Surg.* **1993**, *7*, 633–637. [CrossRef]
42. Yuan, Z.; Lu, Y.; Wei, J.; Wu, J.; Yang, J.; Cai, Z. Abdominal aortic aneurysm: Roles of inflammatory cells. *Front. Immunol.* **2021**, *11*, 3758. [CrossRef]
43. Shimizu, K.; Mitchell, R.N.; Libby, P. Inflammation and cellular immune responses in abdominal aortic aneurysms. *Arterioscler. Thromb. Vasc. Biol.* **2006**, *26*, 987–994. [CrossRef]
44. Michel, J.B.; Martin-Ventura, J.L.; Egido, J.; Sakalihasan, N.; Treska, V.; Lindholt, J.; Allaire, E.; Thorsteinsdottir, U.; Cockerill, G.; Swedenborg, J. Novel aspects of the pathogenesis of aneurysms of the abdominal aorta in humans. *Cardiovasc. Res.* **2011**, *90*, 18–27. [CrossRef]

Article

Goldmann Tonometry and Corneal Biomechanics

Dario Messenio [1], Marco Ferroni [2] and Federica Boschetti [2,3,*]

[1] ASST Fatebenefratelli Sacco, Department of Clinical Sciences, Eye Clinic, University of Milan, 20157 Milan, Italy; dario.messenio@virgilio.it
[2] Department of Chemistry, Materials and Chemical Engineering, "Giulio Natta", Politecnico di Milano, 20133 Milan, Italy; marco.ferroni@polimi.it
[3] IRCCS Istituto Ortopedico Galeazzi, 20161 Milan, Italy
* Correspondence: federica.boschetti@polimi.it

Featured Application: This manuscript presents a critical analysis of intraocular pressure (IOP) measured by Goldmann applanation tonometry (GIOP), the gold standard technique for the measurement of intraocular pressure, compared to measurements obtained by a pressure transducer inserted in the ocular anterior chamber (TIOP). Data showed significant differences between GIOP and TIOP, more evident for softer and thinner corneas and suggest GIOP should be corrected on the basis of corneal biomechanical parameters. This evidence is crucial for the detection and prevention of glaucoma as one of the main causes of irreversible blindness worldwide.

Abstract: Glaucoma is the second cause of irreversible blindness in the world. Intraocular pressure (IOP) is a recognized major risk factor for the development and progression of glaucomatous damage. Goldmann applanation tonometry (GAT) is internationally accepted as the gold standard for the measurement of IOP. The purpose of this study was to search for correlations between Goldmann tonometry and corneal mechanical properties and thickness by means of in vitro tests. IOP was measured by the Goldmann applanation tonometer (GIOP), and by a pressure transducer inserted in the anterior chamber of the eye (TIOP), at increasing pressure levels by addition of saline solution in the anterior chamber of enucleated pig eyes (n = 49). Mechanical properties were also determined by inflation tests. The GAT underestimated the real measurements made by the pressure transducer, with most common differences in the range 15–28 mmHg. The difference between the two instruments, highlighted by the Bland–Altman test, was confirmed by ANOVA, normality tests, and Mann–Whitney's tests, both on the data arranged for infusions and for the data organized by pressure ranges. Pearson correlation tests revealed a negative correlation between (TIOP-GIOP) and both corneal stiffness and corneal thickness. In conclusion, data obtained showed a discrepancy between GIOP and TIOP more evident for softer and thinner corneas, that is very important for glaucoma detection.

Keywords: Goldmann tonometry; intraocular pressure; glaucoma; inflation tests; pig eyes; corneal stiffness

Citation: Messenio, D.; Ferroni, M.; Boschetti, F. Goldmann Tonometry and Corneal Biomechanics. *Appl. Sci.* **2021**, *11*, 4025. https://doi.org/10.3390/app11094025

Academic Editor: Vasudevan (Vengu) Lakshminarayanan

Received: 12 March 2021
Accepted: 27 April 2021
Published: 28 April 2021

1. Introduction

Glaucoma is a multifactorial optic neuropathy characterized by progressive loss of retinal ganglion cells, the neurodegeneration can also involve the neuronal pathways up to the geniculate body and occipital cortex, resulting in changes in optic disk morphology and visual field defects [1–3]. It is the leading cause of irreversible blindness worldwide, accounting for 8% of vision loss [4,5]. Intraocular pressure (IOP) is recognized as the most important risk factor for the development or progression of glaucomatous damage, and is controlled by the balance between aqueous humor secretion and the eye's outflow facility [6].

An association between increased IOP and the loss of sight in glaucoma has been noted for many centuries, from observations of eye stiffness for cases of continued impaired vision following cataract surgery, through the establishment of the relationship between the IOP and loss of sight. In particular, elevated IOP increases the likelihood of visual filed alterations or scotoma and even complete blindness. For this reason, the IOP evaluation remains the primary measurement in the diagnosis of glaucoma [7]. Ocular hypertension studies also investigated and addressed whether the treatment of elevated IOP prevented or delayed the onset of glaucomatous damage, showing that a decrease of IOP reduced the risk of progression to glaucoma [8]. Furthermore, in recent epidemiological studies it has been verified that a reduction in IOP of only 1 mmHg from baseline leads to a 10% reduction of damage progression, and of the conversion from ocular hypertension (i.e., without damage) to manifest glaucoma [9,10]. Therefore, obtaining a correct measurement of IOP during the treatment of the disease or as screening test for the identification of subjects at risk, appears very important. Together with visual field analysis, IOP is the gold standard for the diagnosis and the correct evaluation of the progression of the disease. Recently, useful software has been developed to analyze optical fibers at the peripapillary and macular levels using optical coherence tomography (OCT) to monitor disease progression [11–13]. Unfortunately, in some glaucoma patients there is a progression of the disease, regardless of an apparently normal IOP. Known as normal tension glaucoma, it is an optic disease in which the IOP before ocular hypotonizing therapy is less than 21 mmHg, apparently normal values [14–16]. Many factors, such as the central corneal thickness (CCT) and biomechanical properties of the cornea, may affect IOP measurement, particularly in patients affected by glaucoma [17–19].

The most accurate method to evaluate the IOP is the direct one, which measures the real (actual) pressure in the eye through the cannulation of the eye, using a pressure transducer placed in direct communication with the anterior chamber. This approach, because of its high invasiveness, is not valid for routine measurements of IOP, and has to be considered only for experimental use or during surgery [20–22]. All other methods of measuring IOP are indirect; the cornea is deformed after application of an external force to the corneal surface. The amount of force necessary to obtain changes in corneal normal conformation is proportional to the pressure inside the bulb. In clinical practice IOP is then calculated—not measured directly—by indirect techniques. The tool that allows to make this measurement is called a tonometer (applanation or non-contact tonometry).

In addition to the Goldmann applanation tonometer (GAT), which will be further discussed below, several types of tonometers are currently available, each with advantages and disadvantages [23]. The non-contact tonometry (NCT) is influenced by tear meniscus height (TMH), as the measured IOP increases with increasing tear film, and by central corneal thickness (CCT) [24–26]. The ocular response analyzer (ORA) uses a jet of air as the applanating force to the apex of the cornea; it can measure corneal hysteresis and the corneal resistance factor and provides higher IOP measurements as compared to GAT [27]. Pneumatonometer is a portable instrument which overestimates IOP at high values and at high CCT [24]. The Icare HOME tonometer is a portable device, contact rebound tonometer. It also has been recently shown to overestimate IOP compared to GAT [28]. The dynamic contour tonometer (DCT) or Pascal tonometer differs from GAT having a concave tip equipped with a tiny piezoelectric sensor: it takes about a hundred measurements per second and also evaluates IOP fluctuations with systemic pressure variation; no statistically significant effect of corneal curvature, astigmatism, axial length, and age on the difference between DCT and intracameral IOP has been detected [29,30]. A recent study has been carried out on an IOP sensor mounted on contact lens, not yet available for clinical use, that can be worn by patients for up to 24 h, also measuring the circadian rhythm. The data show an overnight rise in IOP [31]. Finally, the Corvis tonometer, a rebound tonometer, able to take into account corneal biomechanical characteristics, also has been shown to measure higher IOP values compared to GAT [32]. Although observations outlined above underline that most of the tonometers on the market "overestimate" the IOP measured by the Goldmann tonometer, the Goldmann applanation tonometry (GAT) is internationally considered as the gold standard for IOP measurement in clinical practice without the need to deform appreciably the cornea [33,34]. The functioning of this device is based on the measurement of the force required for the applanation of a specific portion (of 3.06 mm in diameter) of the central cornea, and on the use of this measured force to estimate the value of the internal pressure, on the basis of a calibration procedure which depends on a number of standard parameters. The accuracy of tonometry therefore depends on the mechanical resistance of the eye structure to the applanation, which is in turn influenced by the central corneal thickness, the curvature, and the mechanical properties of the cornea [24,35–38]. Recently, the need to make the IOP measurement faster and more comfortable has led to the development of new techniques [39–41].

Analyzing corneal thickness, some researchers found a statistically significant correlation between corneal thickness and GAT [24,42,43], while many others argued that thicker corneas led to an overestimation of the IOP, and on the contrary thinner corneas led to IOP underestimation [24,35,44–47]. Corneal curvature has been recognized as a parameter that reduces the precision in the measurement of IOP [35,47,48].

In summary, Goldman tonometry appears imprecise. It would be very important to know how the geometrical and mechanical properties of the cornea influence and modify the pressure inside the eye, making more or less accurate the measurement made by tonometry. Noted the presence of several studies concerning the correlation between IOP measurement and corneal thickness [19,41,49,50], only a few suggest that corneal biomechanics may affect IOP measurements [19,35,51], inducing significant errors in the diagnosis of glaucoma. Thus, the purpose of this study was first to determine any difference between IOP measured indirectly by the GAT and directly by cannulation of the eye, and then to investigate the existence of statistically significant correlations between such differences and (i) the corneal thickness; and (ii) the corneal elastic modulus. In order to be able to evaluate the mechanical properties of the corneal tissue and its influence on the tonometry measurement, in vitro tests were performed in pig eyes.

2. Materials and Methods

Swine enucleated eyes collected from a local abattoir (Venegoni Spa, Boffalora Sopra Ticino, Italy) were used for the experiments. This type of eyes is very similar to human ones, both in size and mechanical response, and, unlike them, can be obtained with relative ease. Before experimentation, extraocular muscles and periorbital fat were removed. Eyes were kept in the refrigerator at 4 °C until testing. IOP measurement tests were performed the same day of eye collection, pressurization tests the same day or the day after.

2.1. Tonometry Test

The intraocular pressure detected by the Goldmann applanation tonometer (GIOP) was compared to the intraocular pressure (TIOP) invasively measured by a pressure transducer (140205D, Honeywell, Freeport, IL, USA) in 49 swine eyes. Each eye was clamped in a custom-made Plexiglass eye holder placed in vertical position on the chin rest of the slit lamp with the corneal apex in front of Goldmann tonometer, simulating the clinical measurement of IOP. A needle was cannulated at the level of limbus directly into the anterior chamber of the eye and connected to the transducer by a connecting tube and a stopcock. The transducer in turn was connected to a power supply (KAT 5VD, Kert, Caerano di San Marco (TV), Italy). The pressure transducer signal was measured by a multimeter in mV and then converted in mmHg through a mercury sphygmomanometer. A syringe connected to the stopcock was used to gradually increase intraocular pressure by saline infusion, added in steps of 100 μL, up to 500 μL. While infusing the saline solution, through the inspection system at variable magnifications and the optical system of the slit lamp, an applanation area of 7.35 mm^2 was obtained under the Goldmann tonometer. For each saline infusion, TIOP and GIOP were recorded, until the achievement of a TIOP of about 35 mmHg.

The needle used was chosen with the lowest possible diameter (30 gauge), in order to avoid damage to the corneal tissue, once inserted into the eye anterior chamber. The procedure of inserting the needle was carried out with particular care to avoid corneal injuries that can prevent the measurement of IOP by Goldmann tonometry or cause false assessments. Care was taken to avoid leaks in the connection circuit between the transducer and the eye and to minimize pressure losses in the pipes, by using only one tap and a short connecting pipe. Figure 1 is a schematic representation of the IOP measurement setup.

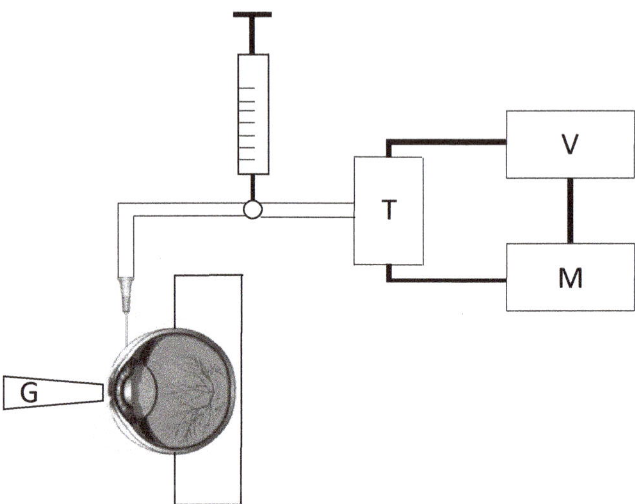

Figure 1. Scheme of the experimental setup used to measure the intraocular pressure, IOP. G: Goldmann applanation tonometer, T: pressure transducer, V: power supply, M: multimeter.

The time between the saline infusion and measurement by Goldmann/transducer was as short as possible (less than one minute) to avoid relaxation effects due to the viscoelastic properties of the ocular tissues [52] that would result in IOP values lower than the actual. During the measurement procedure eyes were kept hydrated by dropping hydrating solution (sodium hyaluronate 0.2%, distilled water). Fluorescein (1 mg fluorescein sodium) was used to visualize the two GAT semi-circles by blue light.

2.2. *Inflation Test*

With the purpose of measuring the corneal stiffness and correlate it to any possible difference between the IOP measured by the two instruments, inflation tests were then performed on the same eyes by cutting the cornea from the eye. A ring of sclera around the cornea was left to allow clamping of the cornea into a custom-made device previously described in [53]. During the inflation tests, each specimen was subjected to three loading-unloading pressure cycles with the pressure ranging from 1.8 to 30 mmHg serving as precondition cycles. Then corneal specimens were subjected to a posterior pressure, P (from 1.8 to 30 mmHg in steps of 2.5 mmHg), induced by a column of NaCl solution to simulate the effect of a growing intraocular pressure. In order to obtain quasi-steady response, we waited three minutes before recording images and applying the subsequent pressure step. Images including the entire profiles of the anterior surface of the cornea were acquired at regular intervals by means of a digital camera (Nikon DS5M, Nital, Moncalieri (TO), Italy), mounted on a stereomicroscope (Nikon SMZ800, Nital, Moncalieri (TO), Italy) and then analyzed by image analysis software (Nikon NIS-Elements D2.20, Nital, Moncalieri (TO), Italy) to track the corneal apex displacement as a function of the applied pressure. The linearized shell theory was applied to data in order to convert apex displacement and pressure in stress and strain, and to calculate the secant modulus, E, for each level of applied pressure, as already described in [53]. The geometrical parameters needed to apply the shell theory, besides the apex displacement, are the average in-plane diameter, *S*, and the radius of curvature, *R*, also evaluated by image analysis, and the corneal thickness at the apex, measured at the end of inflation tests on rectangular corneal strips cut from the tested corneas. The Poisson's coefficient was set to 0.5, consistent with the hypothesis of isotropic material at the basis of the shell theory [54,55]. The experimental setup for the execution of the inflation tests is shown in Figure 2.

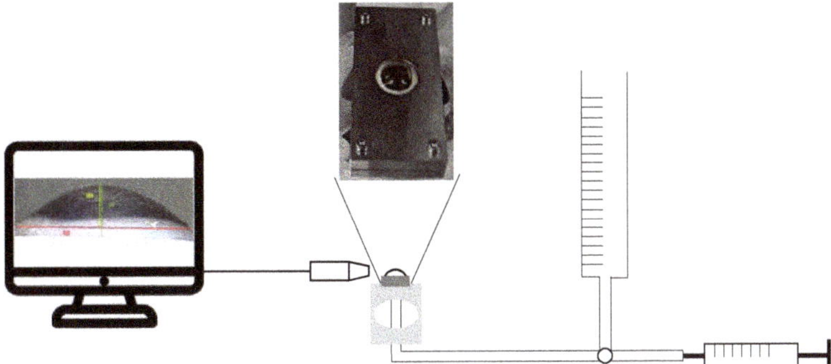

Figure 2. Scheme of the inflation tests experimental setup used to evaluate corneal stiffness.

2.3. Statistical Analysis

All measurements were then analyzed in several ways. In particular, after the evidence of differences between GIOP and TIOP, that we name from now on delta-pressure, a thorough statistical analysis was performed. Shapiro–Wilk test was used to test the normality of our distributions. A two-way ANOVA test was used to test whether volume infusion or instrument type or the pressure level or combinations of the above had any influence on measurements. Pearson correlation test was used to test the presence of a correlation between the delta-pressure and the corneal stiffness, *E*, or the corneal thickness, *CCT*. Finally, Bland–Altman analysis [56,57] was also applied to our data to detect any trend in delta-pressure with the level of IOP. Statistical analysis was performed by Microsoft Excel (Redmond, WA, USA) and Statistic Package for Social Science (SPSS, version 27, Chicago, IL, USA).

Before applying Pearson test to delta-pressure and E, data were properly organized. For each eye, we had the two measurements of IOP following volume infusions, namely GIOP and TIOP, and one value of stiffness, E, for every applied pressure, P, from the inflation tests. We matched the inflation pressure, P, to TIOP, considering pressure ranges of 5 mmHg, and then we obtained for each eye the following data to compare: P, E, delta-pressure. Pearson correlation test was then applied between E and delta-pressure considering the whole set of data for all pressures, P, and 3 sub-sets of data, divided according to the range of *p* values: group 1, from 1 to 12 mmHg, group 2 from 13 to 21 mmHg, group 3 from 22 to 35 mmHg. To perform Pearson tests between delta-pressure and CCT, data were simply matched by considering the thickness of each eye for which the delta-pressure was measured.

3. Results

3.1. Tonometry Results

For each of the 49 pig eyes, we obtained a few couples of measurements of IOP by the two instruments, for a total of 126 measurements.

The values of IOP increase with volume increase for both the instruments, with a trend line for GIOP lower than TIOP, as shown in Figure 3a. Lower GIOP values are also evident from Figure 3b, in which GIOP is plotted versus TIOP, and most of the experimental points are below the first quadrant bisector.

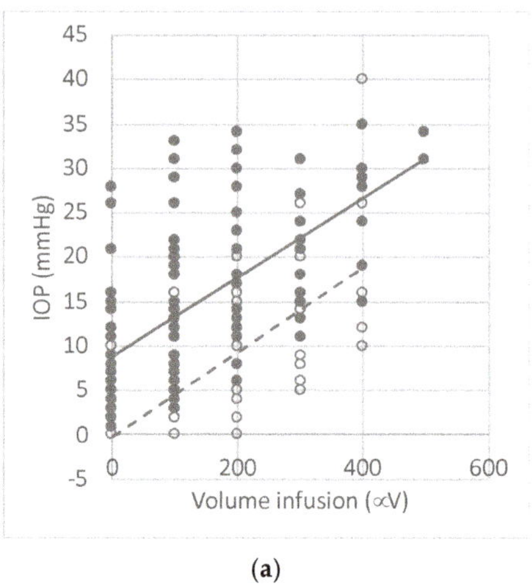

(a)

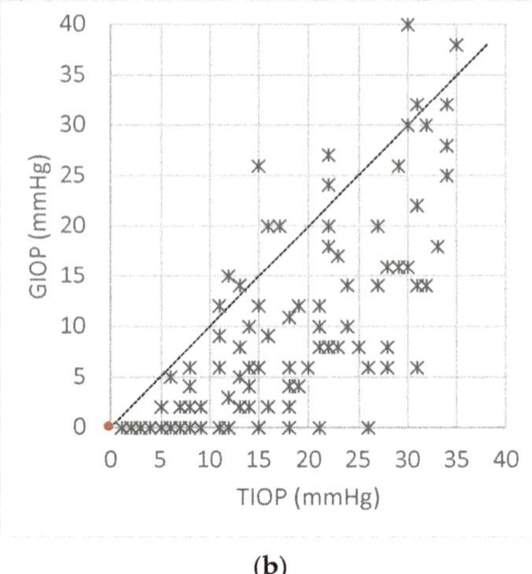

(b)

Figure 3. (**a**) Intraocular pressure, IOP, resulting from discrete volume infusions of 0, 100, 200, 300, 400, and 550 µL. Filled symbols and continuous line are the transducer intraocular pressure, TIOP, data and TIOP trendline, respectively. Empty symbols and dashed line are Goldmann intraocular pressure data, GIOP, and GIOP trendline, respectively. (**b**) GIOP as a function of TIOP measured in the same pig eye. The line represents the first quadrant bisector.

Tables 1 and 2 summarize results of normality test (Shapiro–Wilk) and significance test (*t*-test or Mann–Whitney test) for the five groups of data in Figure 4a. Delta-pressure values of 4–7 mmHg are the most frequent (30%), followed by the intervals 8–11 and 12–15 mmHg (both 20%), as Figure 4b illustrates. *t*-test was applied to normal distributions, Mann–Whitney test otherwise.

Given the presence of consistent differences in IOP measured by the two instruments, we applied a two-way ANOVA test to our data.

Table 1. Shapiro-Wilk test results on the difference between transducer intraocular pressure and Goldmann intraocular pressure (delta-pressure) grouped in five intervals.

Delta-Pressure (mmHg)	Normal Distribution	N
0–7	No	32
8–14	No	37
15–21	No	22
22–28	Yes	18
29–35	Yes	17

Table 2. Significance results between groups of Table 1.

Delta-Pressure (mmHg)	Significance
(0–7) vs. (8–14)	$p < 0.05$ [1]
(8–14) vs. (15–21)	$p < 0.05$ [1]
(15–21) vs. (22–28)	$p < 0.05$ [1]
(22–28) vs. (29–35)	n.s. [2]

[1] Mann-Whitney test, [2] *t*-test. Level of significance $\alpha = 0.05$.

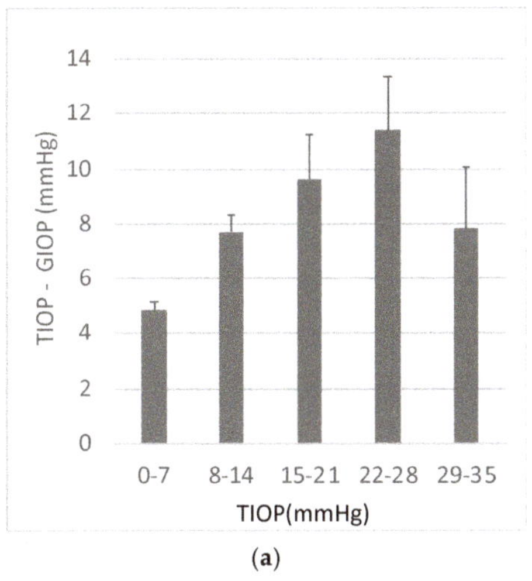

(a)

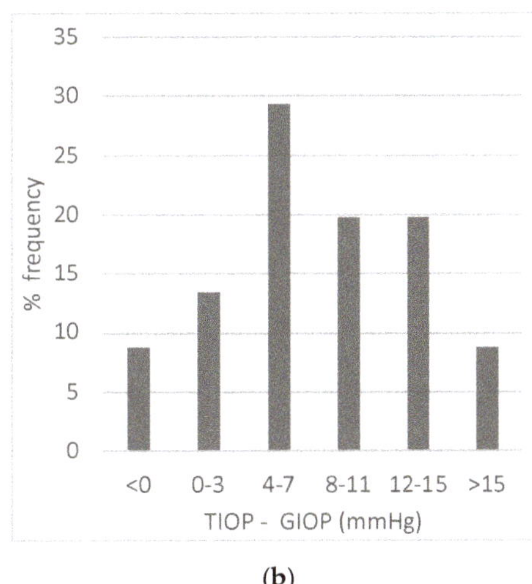

(b)

Figure 4. (a) Distribution of the difference between transducer intraocular pressure and Goldmann intraocular pressure (delta-pressure, TIOP-GIOP) for different intervals of transducer intraocular pressure, TIOP; (**b**) % frequency distribution of different intervals of delta-pressure. Error bars in (**a**) are standard errors.

First, we grouped our data by volume infusion to test whether only volume infusion had an influence on the measured IOP, or the instrument too, or the combination of volume infusion and instrument. Of the total 126 measurements, 78 were relative to 0 mL infusion, 66 to 100 mL, 52 to 200 mL, 30 to 300 mL, 18 to 400 mL, and eight to 500 mL. The *p*-value resulted less than 0.001 for both volume infusion and instrument, whereas the combination of instrument and volume infusion resulted in a *p*-value of 0.51. Therefore, we conclude that both volume infusion and instrument affect the measurement but not the combination of the two.

We also applied two-way ANOVA to data organized by pressure level. Data were divided by five groups identified by the mean pressure of pressure ranges in Table 1, i.e., 4, 11, 18, 25, and 32 mmHg. This time we tested whether the pressure level or the instrument had an influence on IOP measurement. Again, the *p*-value was less than 0.001 for both pressure level and instrument. The combination of instrument and pressure level resulted in a *p*-value of 0.004, so this time also the combination of the two variables had an effect on IOP.

We finally applied Bland–Altman analysis to our data to evaluate more deeply the differences in IOP measured by the two instruments. Figure 5a reports the difference in IOP measurements, delta-pressure, as a function of the average of the two measurements, average. Being most points between the two dashed lines, which represent the confidence lines, the two methods give congruent results, apart from an error. A positive mean value of delta-pressure of 7.8 mmHg is calculated (thick horizontal line in Figure 5a). No trend in delta-pressure with pressure can be observed from Figure 5a. Although delta-pressure for the average lower than 10 mmHg seems lower than for the higher average, Figure 5b shows that such values are on the contrary very high (about 200%) when considered as percentage values (Bland–Altman PI plot, Figure 5b). Figure 5b shows a negative trend in the normalized percentage delta-pressure, confirming a roughly constant delta-pressure vs. the average [58].

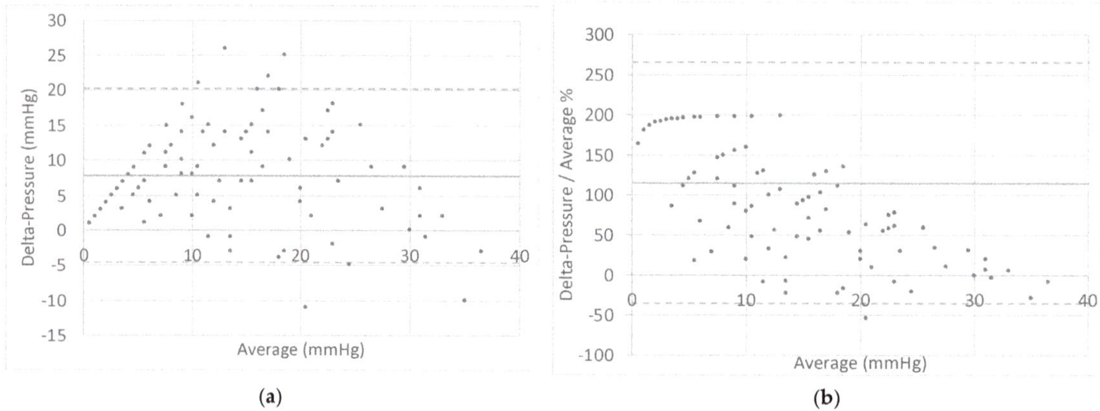

Figure 5. (**a**) Bland–Altman plot: the difference in the measurements of intraocular pressure by the two instruments (delta-pressure) as a function of the average of the two measurements (average); (**b**) Bland–Altman PI plot: Delta-pressure difference is normalized to average and given in percentage.

3.2. Inflation Test Results

From image analysis of corneas in the unstressed state taken during inflation tests, we obtained the geometry data necessary for stress and elastic modulus calculations through the linear shell theory for each of the 49 pig corneas tested. The average geometrical data are listed in Table 3, whereas Table 4 lists the average apex displacement evaluated for each value of inflation pressure. Using such data and the numerical procedure detailed in the Appendix of our previous paper [53], average stress-strain data (Figure 6a), and the distribution of the secant modulus with pressure (Figure 6b) were obtained.

Table 3. Average values from image analysis of corneas during inflation tests. R, radius of curvature, D, in-plane diameter, CCT, thickness at the apex, and H_0, the elevation of the apex in the unstressed condition.

Geometrical Parameter	Value (mm) ± SD
R	8.49 ± 0.49
D	16.98 ± 0.98
CCT	1.41 ± 0.42
H_0	3.53 ± 0.65

Table 4. Apex displacement, w (mm) for every applied pressure, P (mmHg).

P (mmHg)	w (mm) ± SD
3.68	0.26 ± 0.25
7.36	0.35 ± 0.29
11.03	0.40 ± 0.32
14.71	0.43 ± 0.32
18.39	0.47 ± 0.33
22.07	0.50 ± 0.34
25.74	0.53 ± 0.34
29.42	0.55 ± 0.31

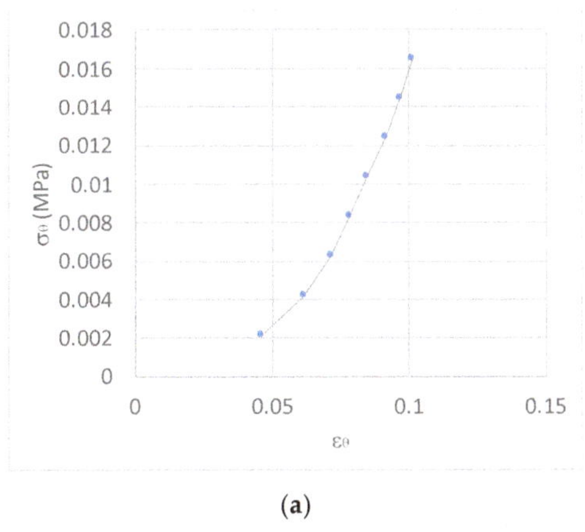

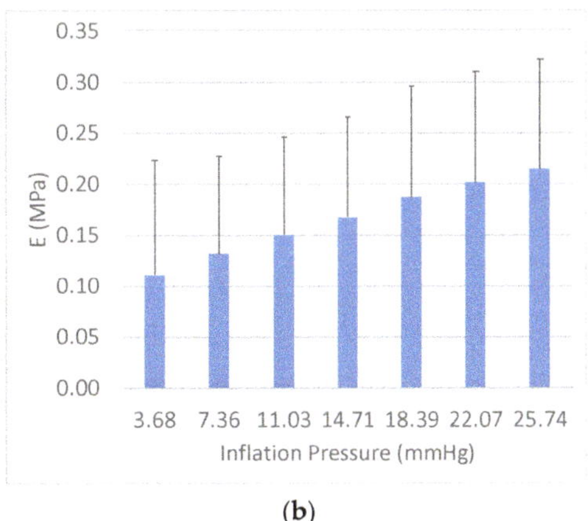

(a) (b)

Figure 6. (**a**) Stress-strain curves for untreated corneas obtained from the average data in Table 3 using the shell theory; (**b**) secant elastic modulus, E, for the applied pressures during inflation tests. Error bars in (**b**) are standard deviations.

3.3. Correlations

Since our hypothesis is that the GAT measurements are affected by the mechanical properties of the cornea, we searched for a correlation between delta-pressure and corneal elastic modulus, *E*, (Pearson correlation test), after organizing our data as described in the materials and methods section. Shapiro–Wilk test applied to delta-pressure resulted in normal distributions for all of the subsets of data presented in Table 5, a condition required for use of Pearson's test. Pearson correlation test results for *E* are shown in Table 5. A weak negative correlation was found for the whole pressure interval, whereas the negative correlation is stronger for pressures in the ranges 13–21 and 22–35 mmHg. A negative Pearson correlation of (−0.38) was found between delta-pressure and the corneal thickness, *CCT*.

Table 5. Pearson correlation indexes between delta-pressure and corneal stiffness, *E*, for different pressure ranges.

IOP Interval (mmHg)	Pearson Correlation
1–35	−0.10
1–12	0.09
13–21	−0.41
22–35	−0.29

4. Discussion

The purpose of the present work was to test the hypothesis that corneal stiffness and thickness are correlated to errors in IOP measured indirectly by GAT, as compared to the true IOP invasively detected using a pressure transducer. The experimental tests were performed on enucleated pig eyes, chosen for their similarity to human eyes both from the anatomical and mechanical point of view [59], also in the studies of glaucoma associated diseases [58,60–62]. ANOVA tests demonstrated that the values of IOP are significantly affected by the instrument used to measure it, therefore deviations of GIOP from TIOP can be attributed to the use of GAT. Although we applied ANOVA test to our two groups of data (GIOP and TIOP) even if none of the two were normally distributed, it should be recalled that ANOVA test remains a valid statistical procedure even under non-normality [63]. GAT in most measurement underestimated IOP measured by the transducer, and the average difference as evaluated by Bland–Altman plots is roughly 8 mmHg. The Bland–Altman plot is the elective method to put in evidence differences in measurements between two instruments. Being such difference roughly constant throughout the whole pressure range, we conclude that GAT underestimated IOP for any level of pressure in our enucleated eyes. Figure 4a shows an increase in delta-pressure with pressure which is not in contrast with Figure 5a, being variables on the x-axis different in the two figures.

Inflation tests confirmed the non-linear mechanical behavior with large data dispersion typical of porcine corneas [53] and other soft biological tissues. The average values of apex displacement reported in Table 4 are very similar to values measured by Bryant and McDonnell [64] in human corneas, but lower than those measured by Elsheikh et al. [65]. The difference may be attributed to corneal viscoelasticity. Our data and those of Bryant and McDonnell [64] were obtained under quasi-static conditions, whereas Elsheikh et al. [65] applied pressures at variable rates from 3.7 to 37.5 mmHg/min. Furthermore, our pressure ranges are much more similar to those of Bryant and McDonnell [64] and in the physiological range.

A weak negative correlation was found between delta-pressure and corneal stiffness, E. Even more importantly, the negative correlation becomes stronger if we look at physiologic or high values of pressure, where it would be important not to underestimate the IOP. For very high values of IOP, the increase in E (Figure 6b) may compensate the errors in GAT (last bar in Figure 4a). In the interval 0–12 mmHg no correlation could be found between delta-pressure and E, we attribute this result to the hypotonic state of eyes. Such pressure ranges are never met in vivo, though, and were measured in our excised eyes which become hypotonic soon after the animal death and need volume infusion to regain tone.

As the GAT is usually calibrated to work with a range of "normal" stiffnesses, our results suggest that GAT introduces larger errors for softer than normal corneas. As a support to this consideration, Susanna et al. [66] reported thin cornea and low corneal hysteresis as main risk factors for glaucomatous visual field progression in eyes with well-controlled IOP. Corneal hysteresis is not E, although it is directly related to it, being defined as the difference between the pressure at which the cornea bends inward during an air jet applanation and the pressure at which it bends out. Sit et al. [67] also measured a lower stiffness in glaucomatous eyes.

A negative correlation was found between delta-pressure and corneal thickness, CCT, meaning that thinner corneas result in greater IOP measurement errors by GAT, and in particular larger underestimation of IOP. The possible influence of corneal thickness in the measurement of IOP was identified and briefly discussed by Goldmann and Schmidt in 1957 [33]. Subsequent studies, performed by Ehlers et al. [44], described in great detail the effect of corneal thickness in the measurement of internal pressure, and the interest in this phenomenon grew again with the advent of refractive surgery. The procedure by which the refractive surgery is performed in fact determines a thinning of the cornea, and was widely discussed, in relation to tonometry, in several experimental works [68]. Orssenigo et al. [45] and Liu et al. [35] found that a high value of corneal thickness leads to an overestimate of IOP, which is line with our findings.

It is known that porcine corneas are thicker (about 1 mm [69]) than human ones (about 600 μm [70]). Average thickness for our porcine corneas was 1.4 mm, meaning that they probably underwent swelling. Nevertheless, since measurements were taken a-posteriori from images of excised strips, a rapid swelling may have occurred as a consequence of the cut of the protective endothelium and epithelium. Although our thickness measurements therefore are affected by an error, the negative correlation with delta-pressures is still valid and suggests that an even larger error would have been measured for thinner corneas.

The accuracy of the GAT depends on the mechanical strength to applanation, which is in turn influenced by central corneal thickness (CCT), by the curvature, and the mechanical properties of cornea and sclera. The measurement of the intraocular pressure by means of tonometry is, therefore, affected by several sources of error, which may lead to an erroneous diagnosis of ocular hypotension or hypertension or glaucoma. In this paper we demonstrated the correlation between such errors and corneal stiffness and corneal thickness.

Our study presents a few limitations. First of all, it was conducted in vitro, so missing several of the in vivo factors which may affect the results. Although this is an obvious drawback, it also allows to perform controlled experiments and to obtain reproducible results. Tests were conducted on pig eyes which are very similar to human eyes but present some differences. In particular, pig corneas are much thicker than human corneas and they may possess different material properties relative to the human corneas. The effect of thickness on the results of the present study has already been discussed above. The interpretation of experimental data on porcine eyes [71,72], showed that the porcine stroma has, on average, mechanical properties very similar to the ones of the human stroma. Tests conducted on human corneas would clearly be of great importance to confirm our results obtained on porcine corneas. Regarding the methods, we evaluated the apex displacement of corneas from 2D images, whereas a 3D strain distribution may have resulted in more precise calculations. We measured the cornea thickness after the experiments on excised strips of the tested corneas. A pachymetry used on corneas during the test would probably give more precise results. Finally, we did not investigate the correlation with the radius of curvature, R, since our pig eyes, originating from animals of the same age of the same facility, were roughly of the same dimension with very low variability in R data (see Table 3).

Future work will be directed to the quantification of possible correlations with the scleral stiffness, since the presence of various types of diseases can easily lead to considerable changes in the elasticity of the scleral tissue [73–75].

In conclusion, glaucoma is a multifactorial opticopathy with a neurodegenerative component also present at the extraocular level [1–3]. Of all the factors to be taken into consideration, at present only IOP certainly plays an important diagnostic role, besides being the target of therapies to treat glaucoma. Obviously, correct measurement and assessment of IOP is essential. Goldmann tonometry has been a gold standard reference for at least 70 years, though with the improvement of technology, it is now being discussed and critically analyzed. In our study, we have clearly shown that several factors, and in particular corneal stiffness and thickness, come into play in the measurement and assessment of ocular tone, and, above all, that the Goldmann tonometer underestimated IOP values with respect to real intraocular pressure especially for softer and thinner than normal corneas. From this point of view, these data could explain, at least in part, why normotensive glaucoma are diagnosed and also why, after medical, par-surgical or surgical hypotonization, glaucomatous disease still progresses in some patients. Further clinical investigation should be carried out in presence of normal GIOP values if the cornea appears particularly thin and soft, whereas the thickness is regularly measured by a pachymeter, at present only the ophthalmologist experience and sensitivity can detect a sub-normal corneal stiffness. The development of a novel instrument, able to measure the force required to flatten the cornea coupled with a biomechanical model would be a great step forward for measuring the real IOP.

Author Contributions: Conceptualization: D.M. and F.B.; validation: F.B. and M.F.; formal analysis: F.B. and M.F.; investigation: F.B. and D.M.; resources: F.B. and D.M.; writing—original draft preparation: D.M. and F.B.; writing—review and editing: F.B., M.F., and D.M.; visualization: F.B. and M.F.; supervision: F.B. All authors have read and agreed to the published version of the manuscript.

Funding: This research received no external funding.

Institutional Review Board Statement: Not applicable. Our study did not involve animals as the experiments were conducted in vitro on pig eyes bought at a local abattoir.

Acknowledgments: Authors are grateful to Francesca Daga, and Ylenia Ferra, for their help in conducting the in vitro experiments.

Conflicts of Interest: The authors declare no conflict of interest.

References

1. Zhang, H.-J.; Mi, X.-S.; So, K.-F. Normal tension glaucoma: From the brain to the eye or the inverse? *Neural Regen. Res.* **2019**, *14*, 1845–1850. [CrossRef]
2. Gupta, N.; Greenberg, G.; de Tilly, L.N.; Gray, B.; Polemidiotis, M.; Yücel, Y.H. Atrophy of the lateral geniculate nucleus in human glaucoma detected by magnetic resonance imaging. *Br. J. Ophthalmol.* **2009**, *93*, 56–60. [CrossRef]
3. Gupta, N.; Ang, L.-C.; Noël de Tilly, L.; Bidaisee, L.; Yücel, Y.H. Human glaucoma and neural degeneration in intracranial optic nerve, lateral geniculate nucleus, and visual cortex. *Br. J. Ophthalmol.* **2006**, *90*, 674–678. [CrossRef] [PubMed]
4. Wang, W.; He, M.; Li, Z.; Huang, W. Epidemiological variations and trends in health burden of glaucoma worldwide. *Acta Ophthalmol.* **2019**, *97*, e349–e355. [CrossRef] [PubMed]
5. Pascolini, D.; Mariotti, S.P. Global estimates of visual impairment: 2010. *Br. J. Ophthalmol.* **2012**, *96*, 614–618. [CrossRef] [PubMed]
6. Japan Glaucoma Society. The Japan Glaucoma Society Guidelines for Glaucoma (3rd Edition). *Nihon Ganka Gakkai Zasshi* **2012**, *116*, 3–46.
7. The Collected Papers of Sir William Bowman, Bart., F.R.S. *Nature* **1893**, *48*, 26. [CrossRef]
8. Kass, M.A.; Heuer, D.K.; Higginbotham, E.J.; Johnson, C.A.; Keltner, J.L.; Miller, J.P.; Parrish, R.K., II; Wilson, M.R.; Gordon, M.O.; Group for the O.H.T.S. The Ocular Hypertension Treatment Study: A Randomized Trial Determines That Topical Ocular Hypotensive Medication Delays or Prevents the Onset of Primary Open-Angle Glaucoma. *Arch. Ophthalmol.* **2002**, *120*, 701–713. [CrossRef]
9. Leskea, M.C.; Heijl, A.; Hyman, L.; Bengtsson, B.; Komaroff, E. Factors for progression and glaucoma treatment: The Early Manifest Glaucoma Trial. *Curr. Opin. Ophthalmol.* **2004**, *15*, 102–106. [CrossRef]
10. Gordon, M.O.; Beiser, J.A.; Brandt, J.D.; Heuer, D.K.; Higginbotham, E.J.; Johnson, C.A.; Keltner, J.L.; Miller, J.P.; Parrish, R.K., II; Wilson, M.R.; et al. The Ocular Hypertension Treatment Study: Baseline Factors That Predict the Onset of Primary Open-Angle Glaucoma. *Arch. Ophthalmol.* **2002**, *120*, 714–720. [CrossRef]
11. Mihaylova, B.; Dimitrova, G. Evaluation of Retinal Nerve Fiber Layer and Inner Macular Layers in Primary Open-Angle Glaucoma with Spectral-Domain Optical Coherence Tomography. *Opt. Nerve* **2019**. [CrossRef]
12. Bussel, I.I.; Wollstein, G.; Schuman, J.S. OCT for glaucoma diagnosis, screening and detection of glaucoma progression. *Br. J. Ophthalmol.* **2014**, *98* (Suppl. S2), ii15–ii19. [CrossRef]
13. Michelessi, M.; Li, T.; Miele, A.; Azuara-Blanco, A.; Qureshi, R.; Virgili, G. Accuracy of optical coherence tomography for diagnosing glaucoma: An overview of systematic reviews. *Br. J. Ophthalmol.* **2021**, *105*, 490–495. [CrossRef] [PubMed]
14. Mallick, J.; Devi, L.; Malik, P.K.; Mallick, J. Update on Normal Tension Glaucoma. *J. Ophthalmic Vis. Res.* **2016**, *11*, 204–208. [CrossRef] [PubMed]
15. Killer, H.E.; Pircher, A. Normal tension glaucoma: Review of current understanding and mechanisms of the pathogenesis. *Eye* **2018**, *32*, 924–930. [CrossRef] [PubMed]
16. Messenio, D.; Marano, G.; Biganzoli, E. Electrophysiological evaluation in normal-tension glaucoma suspects: A pilot study. *J. Model. Ophthalmol.* **2016**, *1*, 9–30.
17. Smedowski, A.; Weglarz, B.; Tarnawska, D.; Kaarniranta, K.; Wylegala, E. Comparison of Three Intraocular Pressure Measurement Methods Including Biomechanical Properties of the Cornea. *Investig. Ophthalmol. Vis. Sci.* **2014**, *55*, 666–673. [CrossRef]
18. Hagishima, M.; Kamiya, K.; Fujimura, F.; Morita, T.; Shoji, N.; Shimizu, K. Effect of corneal astigmatism on intraocular pressure measurement using ocular response analyzer and Goldmann applanation tonometer. *Graefe's Arch. Clin. Exp. Ophthalmol.* **2010**, *248*, 257–262. [CrossRef] [PubMed]
19. Yaoeda, K.; Fukushima, A.; Shirakashi, M.; Fukuchi, T. Comparison of intraocular pressure adjusted by central corneal thickness or corneal biomechanical properties as measured in glaucomatous eyes using noncontact tonometers and the Goldmann applanation tonometer. *Clin. Ophthalmol.* **2016**, *10*, 829–834. [CrossRef]
20. Shim, J.; Kang, S.; Park, Y.; Kim, S.; Go, S.; Lee, E.; Seo, K. Comparative intraocular pressure measurements using three different rebound tonometers through in an ex vivo analysis and clinical trials in canine eyes. *Vet. Ophthalmol.* **2021**, *24* (Suppl. S1), 186–193. [CrossRef]
21. Elsmo, E.J.; Kiland, J.A.; Kaufman, P.L.; McLellan, G.J. Evaluation of rebound tonometry in non-human primates. *Exp. Eye Res.* **2011**, *92*, 268–273. [CrossRef]
22. McCafferty, S.; Levine, J.; Schwiegerling, J.; Enikov, E.T. Goldmann and error correcting tonometry prisms compared to intracameral pressure. *BMC Ophthalmol.* **2018**, *18*, 2. [CrossRef] [PubMed]
23. Aziz, K.; Friedman, D.S. Tonometers—Which one should I use? *Eye* **2018**, *32*, 931–937. [CrossRef] [PubMed]
24. Tonnu, P.-A.; Ho, T.; Newson, T.; El Sheikh, A.; Sharma, K.; White, E.; Bunce, C.; Garway-Heath, D. The influence of central corneal thickness and age on intraocular pressure measured by pneumotonometry, non-contact tonometry, the Tono-Pen XL, and Goldmann applanation tonometry. *Br. J. Ophthalmol.* **2005**, *89*, 851–854. [CrossRef]
25. Yilmaz, I.; Altan, C.; Aygit, E.D.; Alagoz, C.; Baz, O.; Ahmet, S.; Urvasizoglu, S.; Yasa, D.; Demirok, A. Comparison of three methods of tonometry in normal subjects: Goldmann applanation tonometer, non-contact airpuff tonometer, and Tono-Pen XL. *Clin. Ophthalmol.* **2014**, *8*, 1069–1074. [CrossRef] [PubMed]
26. Seol, B.R.; Kang, T.G.; Gu, B. Intraocular pressure according to different types of tonometry (non-contact and Goldmann applanation) in patients with different degrees of bilateral tearing. *PLoS ONE* **2019**, *14*, e0222652. [CrossRef]

27. Kynigopoulos, M.; Schlote, T.; Kotecha, A.; Tzamalis, A.; Pajic, B.; Haefliger, I. Repeatability of intraocular pressure and corneal biomechanical properties measurements by the ocular response analyser. *Klin. Monbl. Augenheilkd.* **2008**, *225*, 357–360. [CrossRef]
28. Takagi, D.; Sawada, A.; Yamamoto, T. Evaluation of a New Rebound Self-tonometer, Icare HOME: Comparison With Goldmann Applanation Tonometer. *J. Glaucoma* **2017**, *26*, 613–618. [CrossRef]
29. Francis, B.A.; Hsieh, A.; Lai, M.-Y.; Chopra, V.; Pena, F.; Azen, S.; Varma, R. Effects of corneal thickness, corneal curvature, and intraocular pressure level on Goldmann applanation tonometry and dynamic contour tonometry. *Ophthalmology* **2007**, *114*, 20–26. [CrossRef]
30. Bochmann, F.; Kaufmann, C.; Thiel, M.A. Dynamic contour tonometry versus Goldmann applanation tonometry: Challenging the gold standard. *Expert Rev. Ophthalmol.* **2010**, *5*, 743–749. [CrossRef]
31. Posarelli, C.; Ortenzio, P.; Ferreras, A.; Toro, M.D.; Passani, A.; Loiudice, P.; Oddone, F.; Casini, G.; Figus, M. Twenty-Four-Hour Contact Lens Sensor Monitoring of Aqueous Humor Dynamics in Surgically or Medically Treated Glaucoma Patients. *J. Ophthalmol.* **2019**, *2019*, 9890831. [CrossRef] [PubMed]
32. Hong, J.; Xu, J.; Wei, A.; Deng, S.X.; Cui, X.; Yu, X.; Sun, X. A new tonometer—the Corvis ST tonometer: Clinical comparison with noncontact and Goldmann applanation tonometers. *Investig. Ophthalmol. Vis. Sci.* **2013**, *54*, 659–665. [CrossRef] [PubMed]
33. Goldmann, H.; Schmidt, T. Über Applanationstonometrie. *Ophthalmologica* **1957**, *134*, 221–242. [CrossRef] [PubMed]
34. Cook, J.A.; Botello, A.P.; Elders, A.; Fathi Ali, A.; Azuara-Blanco, A.; Fraser, C.; McCormack, K.; Margaret Burr, J. Systematic Review of the Agreement of Tonometers with Goldmann Applanation Tonometry. *Ophthalmology* **2012**, *119*, 1552–1557. [CrossRef] [PubMed]
35. Liu, J.; Roberts, C.J. Influence of corneal biomechanical properties on intraocular pressure measurement: Quantitative analysis. *J. Cataract Refract. Surg.* **2005**, *31*, 146–155. [CrossRef]
36. Kaushik, S.; Pandav, S.S.; Banger, A.; Aggarwal, K.; Gupta, A. Relationship between corneal biomechanical properties, central corneal thickness, and intraocular pressure across the spectrum of glaucoma. *Am. J. Ophthalmol.* **2012**, *153*, 840–849.e2. [CrossRef]
37. Costin, B.R.; Fleming, G.P.; Weber, P.A.; Mahmoud, A.M.; Roberts, C.J. Corneal biomechanical properties affect Goldmann applanation tonometry in primary open-angle glaucoma. *J. Glaucoma* **2014**, *23*, 69–74. [CrossRef] [PubMed]
38. Andreanos, K.; Koutsandrea, C.; Papaconstantinou, D.; Diagourtas, A.; Kotoulas, A.; Dimitrakas, P.; Moschos, M.M. Comparison of Goldmann applanation tonometry and Pascal dynamic contour tonometry in relation to central corneal thickness and corneal curvature. *Clin. Ophthalmol.* **2016**, *10*, 2477–2484. [CrossRef] [PubMed]
39. Tejwani, S.; Dinakaran, S.; Joshi, A.; Shetty, R.; Sinha Roy, A. A cross-sectional study to compare intraocular pressure measurement by sequential use of Goldman applanation tonometry, dynamic contour tonometry, ocular response analyzer, and Corvis ST. *Indian J. Ophthalmol.* **2015**, *63*, 815–820. [CrossRef] [PubMed]
40. Oncel, B.; Dinc, U.A.; Orge, F.; Yalvaç, B.I. Comparison of IOP measurement by ocular response analyzer, dynamic contour, Goldmann applanation, and noncontact tonometry. *Eur. J. Ophthalmol.* **2009**, *19*, 936–941. [CrossRef] [PubMed]
41. Chen, M.; Zhang, L.; Xu, J.; Chen, X.; Gu, Y.; Ren, Y.; Wang, K. Comparability of three intraocular pressure measurement: iCare pro rebound, non-contact and Goldmann applanation tonometry in different IOP group. *BMC Ophthalmol.* **2019**, *19*, 225. [CrossRef] [PubMed]
42. Feltgen, N.; Leifert, D.; Funk, J. Correlation between central corneal thickness, applanation tonometry, and direct intracameral IOP readings. *Br. J. Ophthalmol.* **2001**, *85*, 85–87. [CrossRef] [PubMed]
43. Shah, S.; Spedding, C.; Bhojwani, R.; Kwartz, J.; Henson, D.; McLeod, D. Assessment of the diurnal variation in central corneal thickness and intraocular pressure for patients with suspected glaucoma. *Ophthalmology* **2000**, *107*, 1191–1193. [CrossRef]
44. Ehlers, N.; Bramsen, T.; Sperling, S. Applanation Tonometry and Central Corneal Thickness. *Acta Ophthalmol.* **1975**, *53*, 34–43. [CrossRef] [PubMed]
45. Orssengo, G.J.; Pye, D.C. Determination of the True Intraocular Pressure and Modulus of Elasticity of the Human Cornea in vivo. *Bull. Math. Biol.* **1999**, *61*, 551–572. [CrossRef]
46. Gunvant, P.; Baskaran, M.; Vijaya, L.; Joseph, I.S.; Watkins, R.J.; Nallapothula, M.; Broadway, D.C.; O'Leary, D.J. Effect of corneal parameters on measurements using the pulsatile ocular blood flow tonograph and Goldmann applanation tonometer. *Br. J. Ophthalmol.* **2004**, *88*, 518–522. [CrossRef]
47. Weinreb, R.N.; Brandt, J.D.; Garway-Heath, D.; Medeiros, F. *Intraocular Pressure*; Kugler Publications: Amsterdam, The Netherlands, 2007; Volume 4, ISBN 9789062992133.
48. Mark, H.H. Corneal curvature in applanation tonometry. *Am. J. Ophthalmol.* **1973**, *76*, 223–224. [CrossRef]
49. Nemesure, B.; Wu, S.-Y.; Hennis, A.; Leske, M.C.; Group for the B.E.S. Corneal Thickness and Intraocular Pressure in the Barbados Eye Studies. *Arch. Ophthalmol.* **2003**, *121*, 240–244. [CrossRef]
50. Mansoori, T.; Balakrishna, N. Effect of central corneal thickness on intraocular pressure and comparison of Topcon CT-80 non-contact tonometry with Goldmann applanation tonometry. *Clin. Exp. Optom.* **2018**, *101*, 206–212. [CrossRef]
51. Vinciguerra, R.; Rehman, S.; Vallabh, N.A.; Batterbury, M.; Czanner, G.; Choudhary, A.; Cheeseman, R.; Elsheikh, A.; Willoughby, C.E. Corneal biomechanics and biomechanically corrected intraocular pressure in primary open-angle glaucoma, ocular hypertension and controls. *Br. J. Ophthalmol.* **2020**, *104*, 121–126. [CrossRef]
52. Langham, M.E.; Eisenlohr, J.E. A manometric study of the rate of fall of the intraocular pressure in the living and dead eyes of human subjects. *Investig. Ophthalmol.* **1963**, *2*, 72–82.

53. Boschetti, F.; Triacca, V.; Spinelli, L.; Pandolfi, A. Mechanical Characterization of Porcine Corneas. *J. Biomech. Eng.* **2012**, *134*, 031003. [CrossRef] [PubMed]
54. Pinsky, P.M.; van der Heide, D.; Chernyak, D. Computational modeling of mechanical anisotropy in the cornea and sclera. *J. Cataract Refract. Surg.* **2005**, *31*, 136–145. [CrossRef]
55. Elsheikh, A.; Alhasso, D.; Rama, P. Assessment of the epithelium's contribution to corneal biomechanics. *Exp. Eye Res.* **2008**, *86*, 445–451. [CrossRef] [PubMed]
56. Martin Bland, J.; Altman, D.G. Statistical Methods for Assessing Agreement Between Two Methods of Clinical Measurement. *Lancet* **1986**, *327*, 307–310. [CrossRef]
57. Giavarina, D. Understanding Bland Altman analysis. *Biochem. Med.* **2015**, *25*, 141–151. [CrossRef]
58. Brunette, I.; Rosolen, S.G.; Carrier, M.; Abderrahman, M.; Nada, O.; Germain, L.; Proulx, S. Comparison of the pig and feline models for full thickness corneal transplantation. *Vet. Ophthalmol.* **2011**, *14*, 365–377. [CrossRef] [PubMed]
59. Olsen, T.W.; Sanderson, S.; Feng, X.; Hubbard, W.C. Porcine Sclera: Thickness and Surface Area. *Investig. Ophthalmol. Vis. Sci.* **2002**, *43*, 2529–2532.
60. Subasinghe, S.K.; Ogbuehi, K.C.; Mitchell, L.; Dias, G.J. Animal model with structural similarity to human corneal collagen fibrillar arrangement. *Anat. Sci. Int.* **2021**, *96*, 286–293. [CrossRef]
61. Fernandez-Bueno, I.; Pastor, J.C.; Gayoso, M.J.; Alcalde, I.; Garcia, M.T. Müller and macrophage-like cell interactions in an organotypic culture of porcine neuroretina. *Mol. Vis.* **2008**, *14*, 2148–2156.
62. Ruiz-Ederra, J.; García, M.; Hernández, M.; Urcola, H.; Hernández-Barbáchano, E.; Araiz, J.; Vecino, E. The pig eye as a novel model of glaucoma. *Exp. Eye Res.* **2005**, *81*, 561–569. [CrossRef]
63. Blanca, M.J.; Alarcón, R.; Arnau, J.; Bono, R.; Bendayan, R. Non-normal data: Is ANOVA still a valid option? *Psicothema* **2017**, *29*, 552–557. [CrossRef]
64. Bryant, M.R.; McDonnell, P.J. Constitutive laws for biomechanical modeling of refractive surgery. *J. Biomech. Eng.* **1996**, *118*, 473–481. [CrossRef] [PubMed]
65. Elsheikh, A.; Wang, D.; Pye, D. Determination of the modulus of elasticity of the human cornea. *J. Refract. Surg.* **2007**, *23*, 808–818. [CrossRef] [PubMed]
66. Susanna, B.N.; Ogata, N.G.; Jammal, A.A.; Susanna, C.N.; Berchuck, S.I.; Medeiros, F.A. Corneal Biomechanics and Visual Field Progression in Eyes with Seemingly Well-Controlled Intraocular Pressure. *Ophthalmology* **2019**, *126*, 1640–1646. [CrossRef]
67. Sit, A.J.; Kazemi, A.; Zhou, B.; Zhang, X. Comparison of Ocular Biomechanical Properties in Normal and Glaucomatous Eyes Using Ultrasound Surface Wave Elastography. *Investig. Ophthalmol. Vis. Sci.* **2018**, *59*, 1218. [CrossRef]
68. Doughty, M.J.; Laiquzzaman, M.; Müller, A.; Oblak, E.; Button, N.F. Central corneal thickness in European (white) individuals, especially children and the elderly, and assessment of its possible importance in clinical measures of intra-ocular pressure. *Ophthalmic Physiol. Opt.* **2002**, *22*, 491–504. [CrossRef]
69. Hayashi, S.; Osawa, T.; Tohyama, K. Comparative observations on corneas, with special reference to Bowman's layer and Descemet's membrane in mammals and amphibians. *J. Morphol.* **2002**, *254*, 247–258. [CrossRef]
70. Doughty, M.J.; Zaman, M.L. Human corneal thickness and its impact on intraocular pressure measures: A review and meta-analysis approach. *Surv. Ophthalmol.* **2000**, *44*, 367–408. [CrossRef]
71. Pandolfi, A.; Holzapfel, G.A. Three-Dimensional Modeling and Computational Analysis of the Human Cornea Considering Distributed Collagen Fibril Orientations. *J. Biomech. Eng.* **2008**, *130*, 061006. [CrossRef] [PubMed]
72. Pandolfi, A.; Boschetti, F. The influence of the geometry of the porcine cornea on the biomechanical response of inflation tests. *Comput. Methods Biomech. Biomed. Engin.* **2015**, *18*, 64–77. [CrossRef] [PubMed]
73. Sergienko, N.M.; Shargorogska, I. The scleral rigidity of eyes with different refractions. *Graefe's Arch. Clin. Exp. Ophthalmol.* **2012**, *250*, 1009–1012. [CrossRef] [PubMed]
74. Friedman, E.; Ivry, M.; Ebert, E.; Glynn, R.; Gragoudas, E.; Seddon, J. Increased scleral rigidity and age-related macular degeneration. *Ophthalmology* **1989**, *96*, 104–108. [CrossRef]
75. Graebel, W.P.; van Alphen, G.W.H.M. The Elasticity of Sclera and Choroid of the Human Eye, and Its Implications on Scleral Rigidity and Accommodation. *J. Biomech. Eng.* **1977**, *99*, 203–208. [CrossRef]

 applied sciences

MDPI

Review

The Evolution of Fabrication Methods in Human Retina Regeneration

Beatrice Belgio [1,*], Anna Paola Salvetti [2,3], Sara Mantero [1] and Federica Boschetti [1]

1 Department of Chemistry, Materials and Chemical Engineering "Giulio Natta", Politecnico di Milano, 20133 Milan, Italy; sara.mantero@polimi.it (S.M.); federica.boschetti@polimi.it (F.B.)
2 Department of Biomedical and Clinical Science Luigi Sacco, Fatebenefratelli and Sacco Hospital, University of Milan, 20157 Milan, Italy; paola.anna.salvetti@gmail.com
3 Ophthalmology Department, Casa di Cura Beato Palazzolo, 24122 Bergamo, Italy
* Correspondence: beatrice.belgio@polimi.it

Abstract: Optic nerve and retinal diseases such as age-related macular degeneration and inherited retinal dystrophies (IRDs) often cause permanent sight loss. Currently, a limited number of retinal diseases can be treated. Hence, new strategies are needed. Regenerative medicine and especially tissue engineering have recently emerged as promising alternatives to repair retinal degeneration and recover vision. Here, we provide an overview of retinal anatomy and diseases and a comprehensive review of retinal regeneration approaches. In the first part of the review, we present scaffold-free approaches such as gene therapy and cell sheet technology while in the second part, we focus on fabrication techniques to produce a retinal scaffold with a particular emphasis on recent trends and advances in fabrication techniques. To this end, the use of electrospinning, 3D bioprinting and lithography in retinal regeneration was explored.

Keywords: retina; tissue engineering; retina regeneration; biofabrication; 3D bioprinting; electrospinning; ophthalmology

Citation: Belgio, B.; Salvetti, A.P.; Mantero, S.; Boschetti, F. The Evolution of Fabrication Methods in Human Retina Regeneration. *Appl. Sci.* **2021**, *11*, 4102. https://doi.org/10.3390/app11094102

Academic Editor: Rossella Bedini

Received: 28 February 2021
Accepted: 27 April 2021
Published: 30 April 2021

1. Background

The retina, lining the inner surface of the eye's posterior segment, is a thin light-sensing tissue responsible for light absorption, conversion to an electrical signal and transmission to the brain through the optic nerve. The retina comprises multiple layers of different cells as shown in Figure 1 [1]. Photoreceptors are specialized neurons that convert visual stimuli into electrical impulses. The signal is then processed and transmitted from photoreceptors to the brain by neural cells including retinal ganglion cells (RGCs). The photoreceptors collaborate closely with the underlying retinal pigment epithelium (RPE), which is essential for visual function [2–4]. Hence, the dysfunction or degeneration of RPE cells results in the death of photoreceptors and vision loss [5]. The RPE cell basal surface faces Bruch's membrane, a thin (2–4 μm) acellular matrix located between the retina and the choroid. This membrane presents a nanofibrous structure composed of collagen I-V, laminin and fibronectin [6]. Bruch's membrane serves as a physical support for RPE cells and as a barrier regulating the diffusion of biomolecules, nutrients, oxygen, fluids and waste between the retina and the choroidal blood supply [7,8].

Pathologies of the retina and optic nerve represent a leading cause of visual impairment and irreversible blindness in high-income countries [9]. These diseases are usually divided into those caused by mutation in only one gene (monogenic diseases) such as inherited retinal diseases (IRDs) and those caused by mutations in multiple genes and environmental factors (polygenic and/or multifactorial diseases) such as age-related macular degeneration (AMD) and glaucoma [10–14]. As the retina is composed of neuronal highly specialized cells with a limited healing potential, regenerative strategies are necessary to reverse vision loss caused by these diseases.

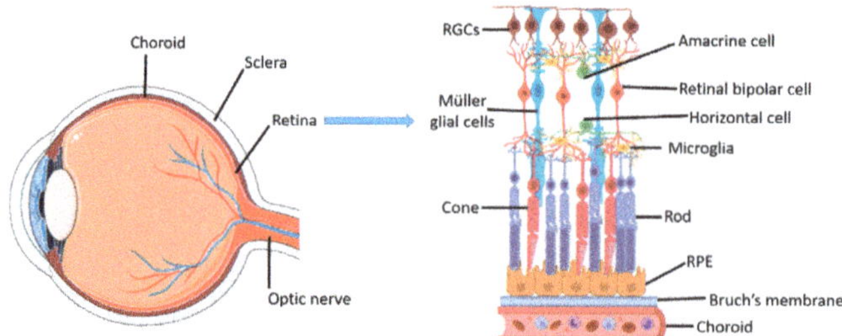

Figure 1. Representation of a cross-section of a human eye and a detail of the multilayer organization of the retina, Bruch's membrane and choroid. The retinal layers include a retinal pigment epithelium (RPE), cone and rod photoreceptors, horizontal cells, bipolar cells, amacrine cells, microglia and Müller glial cells and ganglion cells (RGCs).

This review aims to provide a current account of the developments in retinal regeneration. Retinal regeneration includes tissue engineering approaches and gene- and cell-based therapies, which we refer to as scaffold-free approaches and that are briefly presented in Section 2. Tissue engineering strategies combining cells, a scaffold and chemical cues offer several solutions that are described in Section 3. In particular, we will focus on the regeneration techniques of RGCs, RPE cells along with Bruch's membrane and photoreceptors as these are the cells mainly involved in the majority of pathologies inducing blindness.

2. Scaffold-Free Approaches

Gene therapy has been proven to restore vision by replacing absent or abnormal genes causing monogenic retinal diseases [15–17]. Viral vectors and non-viral gene nanocarriers have been used to deliver specific genes to target cells [15]. For instance, for RPE65-associated IRDs, nowadays there is an FDA (Food and Drug Administration) and EMA (European Medicine Agency)-approved gene therapy (Luxturna (voretigene neparvovec-rzyl), Spark Therapeutics) based on the delivery of a functional copy of the RPE65 gene into RPE cells [17]. Therapies targeting monogenic retinal dystrophies are the most promising. Gene augmentation therapy, however, has its own limitations. Only the recessive forms where the mutations cause a loss of function, also called haploinsufficiency, such that the resulting protein is too little or absent can be currently treated with gene therapy [18].

Another scaffold-free strategy explored to regenerate damaged cells or replace dead cells is cell therapy. Healthy cells can be injected as cell suspension or transplanted as sheets to replace pathological cells, thus preventing further degeneration and improving visual function. For instance, as a possible treatment for glaucoma, the transplantation of RGCs derived from human pluripotent stem cells and from human Müller glia cells was investigated. However, difficulties remain in integrating the transplanted RGCs into the complex neurological network of the host retina [19]. Cell transplantation in animal models has been experimented also to treat AMD. Capela et al. reported that a subretinal injection of human central nervous system stem cells in pigmented dystrophic rats enhanced the proliferation of the host RPE cells, providing a new mechanism for RPE regeneration and thus preserving the viability of photoreceptors [20]. Lu et al. demonstrated that the embryonic stem cell-derived RPE implanted in a pathological mouse model was able to restore visual function in the short term. After 90 days, the visual acuity started decreasing [21]. This reduced efficacy could be due to the lack of interaction between the transplanted cells and Bruch's membrane, thereby not forming a functional monolayer. Phase I/II clinical trials were carried out to determine the primary endpoints of safety and tolerability of a subretinal injection of a cell suspension [22]. No serious adverse effect was encountered; however, concerns about integration efficiency and long term cell survival were raised. A possible approach to overcome this issue is represented by cell sheet engineering [23,24]. Cell sheet engineering is based on harvesting a sheet of cells along with their extracellular matrix (ECM) without the use of enzymes [25]. Harvesting cells is performed with thermoresponsive coatings that enable reversible cell detachment by switching their surface hydrophobicity. Such an approach may allow the formation of an intact RPE cell monolayer to be transplanted. Furthermore, the presence of an ECM should improve the attachment to the host tissue once implanted [26]. However, the use of cell sheet engineering for the RPE is limited by the insufficient amount of the ECM secreted by these cells [18]. In addition, when multiple structures are involved in the pathology, as in AMD, cell therapy cannot be effective in the long term. For example, Bruch's membrane is also compromised in AMD; therefore, it needs to be replaced along with the RPE monolayer. Alternative strategies are needed to fabricate a mechanically strong tissue that promotes cell attachment and survival while maintaining cell functionality in the long term.

3. Tissue Engineering Approach

Tissue engineering has been commonly employed in biomedical applications [27–30]. This approach relies on the use of scaffolds as support systems for adherent cells. Scaffolds should provide a proper in vivo-like microenvironment for tissue regeneration.

Autologous Bruch's membrane explants or its constituent layers were first used as a scaffold for RPE cells to regenerate the outer retina in patients with AMD. According to early studies, the degree of structural support for cell attachment decreases with a decreasing concentration of proteins contained in the RPE basal lamina such as specific laminins [31–35]. To improve the RPE attachment, Bruch's membrane explants were coated with fibronectin, laminin and vitronectin [36–38]. The cell attachment was also enhanced by seeding the RPE cells onto Bruch's membrane explants coated with an ECM previously secreted by corneal endothelial cells [39]. However, the inter-donor variability and limited availability of a native Bruch's membrane have urged researchers to use artificial scaffolds for clinical applications.

The fabrication of artificial scaffolds offers a valid alternative to native tissues. A variety of three-dimensional (3D) scaffolds have been designed and developed for the skin, the bladder, cartilage, bones and muscles [40–44]. Selecting a scaffold fabrication technique and chemical composition is crucial to reproducing the microarchitecture of a native ECM. Here, we provide an overview of the fabrication methods used to engineer a human retina (Figure 2). First, we present the conventional and common methods for retinal scaffold production. We then lay emphasis on the most recent and promising advances in fabrication techniques used for retina tissue engineering.

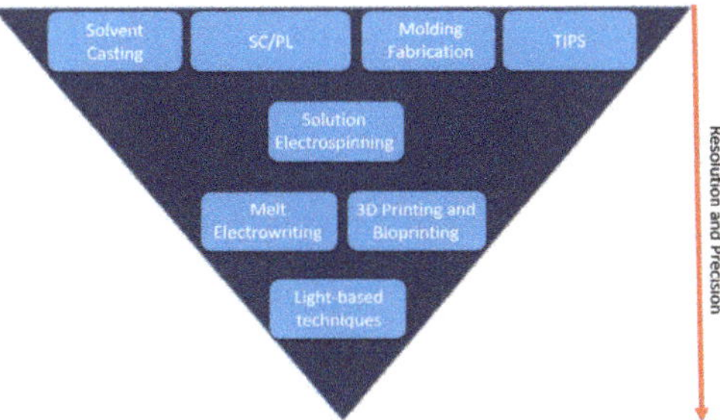

Figure 2. Overview of fabrication methods used for retinal tissue engineering. SC/PL stands for solvent casting-particulate leaching and TIPS for thermally induced phase separation. The red arrow indicates the increase in precision and resolution of the techniques.

3.1. Conventional Fabrication Techniques

Solvent casting techniques have been highly utilized in ocular tissue engineering [45]. They consist of dissolving a polymer in an appropriate solvent followed by casting and solvent evaporation [46]. The resulting scaffold is a uniform and non-porous film. In retinal tissue engineering, these films were seeded with RPE cells to regenerate the RPE layer and replace Bruch's membrane for AMD treatment, thus preventing the loss of photoreceptors [47–51]. The prosthetic Bruch's membrane manufactured by solvent casting displayed a similar thickness to the human Bruch's membrane and promoted correct implantation and orientation of the cellular graft in the subretinal space. Moreover, as the degradation of the polymer proceeds, transplanted RPE cells should re-establish interactions with the native Bruch's membrane [47]. To enhance the survival of a functional RPE monolayer in the long term, the film surface was modified [52,53]. Singh et al. showed that silk fibroin scaffolds coated with type I collagen promoted the in vitro development and survival of a functional RPE monolayer for 90 days [52]. Hasirci et al. used oxygen plasma treatment to render a film surface hydrophilic, thus enhancing cell attachment and spreading [53]. However, the film's non-porous structure did not match the open fibrillar structure of the native Bruch's membrane and could prevent nutrient diffusion from the choroid in vivo. To generate a porous scaffold, a solvent casting/particle leaching (SC/PL) method was adopted in biomedical applications [54]. In the SC/PL technique, the solvent, where the polymer is dissolved, contains leachable particles known as porogen [55]. This method allows the production of scaffolds with high porosity (up to 93%) in a simple, inexpensive way [54]. Prosthetic Bruch's membranes for RPE transplantation were successfully prepared using the SC/PL method [56–58]. Moreover, to develop an optimized cell therapy strategy, the influence of the scaffold pore size on the cell differentiation into the retinal precursor has been investigated [59]. However, the low reproducibility limits the use of the SC/PL technique for retinal tissue engineering applications [60].

Over the last decade, hydrogels have received considerable attention as leading candidates for engineering soft tissues such as the retina due to their unique compositional and structural similarities to the ECM [61]. As such, they are highly hydrophilic and biocompatible, thus promoting cell attachment, proliferation and differentiation. Various methods have been used to produce hydrogels depending on the desired structure and application [61]. The most common method of hydrogel preparation involves the crosslinking of the polymer solution poured into a mold of a specific shape. Crosslinking may occur by physical or chemical methods; the type and degree of crosslinking strongly affects the properties of hydrogels such as the elastic modulus, nutrient transport and swelling capacity [62]. In retina tissue engineering, hydrogels have been employed as scaffolds for RPE, photoreceptors and the regeneration of RGCs [63–70]. For instance, implantable hydrogels with encapsulated cells were proposed as promising candidates to treat those blinding diseases caused by the degeneration of the RPE. Hydrogels acted as a cell delivery vehicle providing an appropriate environment for RPE regeneration [63–65]. Marmorstein et al. seeded an induced pluripotent stem cell (iPSC)-derived RPE onto the surface of thin fibrin hydrogels as a support material for RPE transplantation [66]. The findings of this study indicated that fibrin hydrogels could be easily manipulated with surgical tools and degraded rapidly into products recognized and metabolized by the body. IPSC-RPE cells on a fibrin hydrogel were viable and organized into a functional monolayer that could be preserved during and after implantation thanks to the hydrogel support [66]. However, despite their high biocompatibility, hydrogels have low mechanical properties and, hence, they are not suitable as a Bruch's membrane substitute. In fact, hydrogel stiffness ranges from a few Pa to few kPa whereas the elastic modulus of Bruch's membrane is estimated to be around few MPa [5]. Hydrogels were also employed as 3D scaffolds for neural retina tissue engineering to regenerate damaged photoreceptors and RGCs [69,70]. Conjunctiva stem cells encapsulated in fibrin gels differentiated into photoreceptor-like cells, thus providing a new potential therapy for diseases involving photoreceptor injuries [70]. Similarly, retinal ganglion-like cells were differentiated from dental pulp stem cells suspended in fibrin gels [69]. This construct could be transplanted into patients with glaucoma to repair axonal damage and prevent further RGC death. The 3D networks resembled the physiological 3D microenvironment of the neural retina, thus promoting cell viability, differentiation and function when compared with 2D cultures [68–70]. However, the molding fabrication method has a limited microarchitecture and cell distribution controllability.

Another technique that has been explored in retinal tissue engineering is thermally induced phase separation (TIPS). TIPS induces the destabilization of a homogenous solution composed of a polymer dissolved in a solvent [71]. After dissolution, a polymer can be thermodynamically unstable at low temperatures, thus causing a spontaneous separation of the solution into a polymer-rich phase and a solvent-rich phase. Lowering the temperature in a controlled manner causes the solidification of the solvent and subsequently the separation of the polymer from the solvent. The solvent crystals are then removed through extraction, evaporation and sublimation resulting in porous structures [72]. The TIPS method has been largely employed for tissue engineering applications in general, for example, bones and cartilage, due to its ability to produce highly porous scaffolds with interconnected pores [73–75]. Porous membranes, fabricated by a separation phase, were investigated as a scaffold for RPE transplantation [76]. TIPS was also used to fabricate scaffolds for retinal progenitor cells isolated from mice eyes [77]. Seeded scaffolds were tested in vitro and then implanted into models of retinal degeneration. The outcomes of this study indicated that the scaffolds promoted cell differentiation in vitro and cell survival in vivo [77]. The guidance of cell differentiation by the substrate could be exploited to form functional photoreceptors to replace the cells lost due to degenerative diseases. However, the use of the TIPS technique is limited by the low control of pore size and geometry [73].

The retinal scaffolds produced by conventional techniques are listed in Table 1. Due to their own limits, the conventional fabrication methods have been mostly replaced by innovative techniques that hold great promise as potential treatments for retinal degenerative diseases.

Table 1. A summary of scaffolds produced by conventional methods and used for retinal tissue engineering. This table highlights the fabrication technique, the structure of the scaffold and the cell type used. SC/PL: solvent casting-particle leaching; TIPS: thermally induced phase separation; RPE: retinal pigment epithelium; ESC: embryonic stem cells; iPSC: induced pluripotent stem cells; RGC: retinal ganglion cells.

Fabrication Technique	Scaffold Structure	Cell Type	Research Stage	Reference
Solvent casting	Non-porous film	Human fetal RPE cells	In vitro	[47]
Solvent casting	Non-porous film	Human cell line ARPE-19	In vitro and in vivo (rabbit)	[48]
Solvent casting	Non-porous film	Human cell line ARPE-19	In vitro and in vivo (rabbit)	[49]
Solvent casting	Non-porous film	Human cell line ARPE-19	In vitro	[50]
Solvent casting	Non-porous film	Human RPE cells	In vitro	[51]
Solvent casting	Non-porous film	Human iPSC-RPE cells	In vitro	[52]
Solvent casting	Non-porous film	Human cell line D407	In vitro	[53]
SC/PL	Porous membrane	Human cell line ARPE-19	In vitro	[56]
SC/PL	Porous membrane	Human fetal RPE cells	In vitro	[57]
SC/PL	Porous membrane	Human ESC-RPE cells	In vitro	[58]
SC/PL	Porous membrane	Human or pig RPE	In vitro	[59]
Silicon Mold	3D hydrogel with cells encapsulated	Rabbit RPE cells	In vitro	[63]
Petri dish Mold	3D hydrogel with cells encapsulated	Human cell line ARPE-19	In vitro	[64]
Mold	3D hydrogel with cells encapsulated	Human iPSC- and ESC-derived embryoid bodies	In vitro	[65]
Custom or well plates mold	Thin hydrogel	Human iPSC-RPE cells	In vitro	[66]
Custom mold	Thin gel film	Human cell line ARPE-19	In vitro	[67]
Mold	3D hydrogel	Rat RGCs and amacrine cells	In vitro	[68]
Well plate mold	3D hydrogel with cells encapsulated	Rat dental pulp stem cells	In vitro	[69]
Well plate mold	3D hydrogel with cells encapsulated	Human conjunctiva mesenchymal stem cells	In vitro	[70]
TIPS	Porous membrane	Human cell line ARPE-19	In vitro	[76]
TIPS	Porous membrane	Rat retinal progenitor cells	In vitro and in vivo (rats)	[77]

3.2. Electrospinning

The electrospinning process has been suggested as a promising technique to fabricate a prosthetic Bruch's membrane as it is able to recapitulate the nanofibrous structure of a native Bruch's membrane [78]. Indeed, the electrospinning technique is able to produce 3D thin nanofibrous membranes by using natural and synthetic polymers [79]. These fibrillar networks are highly permeable for solutes, thus facilitating cell adhesion and proliferation [79]. The electrospinning setup involves a high-voltage supply, a capillary tube/syringe with a needle, a syringe pump and a collector. The high-voltage supply applies a positive charge to a polymeric solution (solution electrospinning) or melt (melt electrospinning) that is extruded from the needle forming a jet. The jet becomes unstable and thin and forms fibers while the solvent evaporates. The fibers are deposited onto a collector [80]. The morphological properties of electrospun membranes such as thickness, fiber size and orientation can be tuned simply by changing the specific parameters of the electrospinning process such as the polymer concentration, solution flow rate and collector distance [81]. Many studies have reported the fabrication of electrospun scaffolds for vascular, cardiac and neural tissue engineering [82–84].

Solution electrospinning has been utilized to produce a prosthetic Bruch's membrane to engineer an RPE layer. The hypothesis is that the RPE monolayer engineered on thin electrospun membranes could become an effective therapy to cure blindness and the deficiencies associated with RPE and Bruch's membrane degeneration. Synthetic and natural polymers including poly(lactic acid) (PLA), poly(ε-caprolactone) (PCL) and poly(lactic-co-glycolic acid) (PLGA), silk fibroin and silk fibroin-PCL-gelatin were used to fabricate electrospun Bruch's membrane-like scaffolds [85–100]. Typically, a blend of natural and synthetic materials is electrospun as synthetic polymers exhibit good mechanical properties and a controlled degradation rate while pure naturally-derived polymers promote cell adhesion and proliferation. Warnke et al. successfully produced ultrathin nanofibrous scaffolds based on collagen and PLGA that closely imitated the structure of a native Bruch's membrane [85]. Human RPE cells seeded onto these scaffolds formed a functional monolayer with a typical cobblestone morphology and abundant sheet-like microvilli on their apical surfaces; however, no results of the scaffold mechanics and permeability were reported [85]. In our previous paper, we fabricated a Bruch's membrane-like membrane composed of a blend of *Bombyx mori* silk fibroin and PCL to study the pathological mechanisms of AMD in a 3D in vitro model [100]. The resulting scaffolds showed similarities with a human Bruch's membrane with regard to the architecture, permeability and mechanical properties [100]. These in vitro studies indicated the feasibility of using electrospun membranes as a prosthetic Bruch's membrane on which a functional RPE monolayer is formed. As such, RPE patches, composed of electrospun scaffolds previously seeded with RPE cells, have been investigated in vivo. The RPE patches were proven to be biocompatible when implanted in animal models showing no adverse reactions [86,89,92,95,98]. Sharma et al. also evaluated the functionality of clinical-grade iPSC-RPE patches in rats and in a porcine laser-injured model. According to the findings, the use of the scaffold improved the patch integration and efficacy over the cell suspension [95]. In fact, an increased photoreceptor preservation was encountered in animals transplanted with the patch in respect to those injected with the cell suspension. This study has led to a phase I/II clinical trial (NCT04339764) based on patch subretinal transplantation in patients suffering from advanced stages of dry AMD.

Electrospun scaffolds have been employed also in neural retina engineering to guide the growth of RGC axons through the control of fiber orientation [101–103]. This approach could benefit patients suffering from glaucoma and optic nerve diseases. The transplantation of functional RGCs onto a suitable support combined with treatments that promote dendritic integration may regenerate axons and help recreate a healthy axonal transport. Radially oriented scaffolds have been investigated for the transplantation of RGCs as they mimic the radial axon pattern [101]. It was observed that radially oriented scaffolds increased the survival of RGCs while preserving the cellular electrophysiological function. Moreover, these scaffolds promoted RGC axonal integration with the host retinal nerve fibers in retinal rat explants whereas RGCs transplanted directly onto explants grew axons in a random pattern [101]. Soleimani et al. compared radially and randomly oriented scaffolds to regenerate photoreceptors. They found that the expression of rod photoreceptor-specific genes increased when stem cells were differentiated on randomly oriented nanofibers [104].

Electrospun scaffolds for retina applications are summarized in Table 2.

Table 2. A summary of retinal scaffolds fabricated by electrospinning. This table highlights the fabrication technique, the structure of the scaffold and the cell type used. RPE: retinal pigment epithelium; ESC: embryonic stem cells; iPSC: induced pluripotent stem cells; RGC: retinal ganglion cells.

Fabrication Technique	Scaffold Structure	Cell Type	Research Stage	Reference
Electrospinning	Ultrathin random nanofibrous membrane	Human RPE cells	In vitro	[85]
Electrospinning	Ultrathin random nanofibrous membrane	Human RPE cells	In vitro and in vivo (rabbit)	[86]
Electrospinning	Random nanofibrous membrane	Rat retinal progenitor cells	In vitro	[87]
Electrospinning	Random nanofibrous membrane	Porcine RPE cells	In vitro and ex vivo (pig)	[88]
Electrospinning	Random nanofibrous membrane	Human RPE cells	In vitro and in vivo (rat)	[89]
Electrospinning	Random nanofibrous membrane	Human fetal and adult RPE cells	In vitro	[90]
Electrospinning	Random nanofibrous membrane	Human RPE cells	In vitro	[91]
Electrospinning	Random nanofibrous membrane	Human fetal RPE cells	In vitro and in vivo (rabbit)	[92]
Electrospinning	Random nanofibrous membrane	Human RPE cells	In vitro	[93]
Electrospinning	Random nanofibrous membrane	Human ESC-RPE cells	In vitro	[94]
Electrospinning	Random nanofibrous membrane	Human iPSC-RPE cells	In vitro and in vivo (rat/porcine)	[95]
Electrospinning	Nanofibrous membrane	Human ESC-RPE/bovine RPE cells	In vitro	[96]
Electrospinning	Random nanofibrous membrane	Human RPE cells	In vitro	[97]
Electrospinning	Random nanofibrous membrane	Human cell line ARPE-19/MIO-M1	In vitro and in vivo (rat)	[98]
Electrospinning	Random nanofibrous membrane	Rat retinal progenitor cells	In vitro	[99]
Electrospinning	Random nanofibrous membrane	Human cell line ARPE-19	In vitro	[100]
Electrospinning	Aligned nanofibrous membrane	Rat RGCs	In vitro and ex vivo (rat)	[101]
Electrospinning	Aligned nanofibrous membrane	Rat RGCs	In vitro	[102]
Electrospinning	Random nanofibrous membrane	Human iPSC-RGCs	In vitro and in vivo (rabbit/monkey)	[103]
Electrospinning	Random + aligned nanofibrous membranes	Human conjunctiva stem cells	In vitro	[104]

Despite the advantages, the electrospinning technique has a limited microarchitecture controllability. Therefore, to overcome this drawback, melt electrowriting (MEW) has been recently developed [105]. MEW allows the controlled deposition of a polymer melt fiber starting from a digital model, thus combining melt electrospinning and additive manufacturing principles. To the best of our knowledge, this innovative method has been used in vascular, bone and skin tissue engineering but no studies on a retinal application are yet present [106–108].

3.3. Lithography

Lithography is a technique that allows the formation of precise and complex 2D and 3D microarchitectures and nanoarchitectures. In the field of biomedicine, photolithography has offered a promising alternative to a conventional fabrication method [109]. Photolithography is a photon-based technique that exploits light to project a mask into a photosensitive emulsion (photoresist) coated onto a substrate. The master fabricated by photolithography can be used to create a polymer negative mold typically of polydimethylsiloxane (PDMS) for a polymeric scaffold fabrication [109]. Retinal progenitor cells were successfully seeded onto structures fabricated through photolithography [110–112]. The microfabricated topography enhanced the attachment, organization and differentiation of the progenitor cells into photoreceptor-like cells [111]. Hence, these structures could be used for photoreceptor replacement in the treatment of photoreceptor degeneration. Photolithography might be followed by wet and ion etching. For instance, Lu et al. developed and tested in vitro a parylene-C membrane artificial Bruch's membrane for an RPE cell culture as a potential treatment for dry AMD [8,113]. In vivo studies demonstrated the safety and potential of a parylene membrane as an RPE scaffold [8]. In particular, the implantation of the membrane previously seeded with cells was compared with the injection of a cell suspension. The results showed that cell survival was greater in animals that were transplanted with the cell membrane patch than those that received the cell suspension. Moreover, when injected, cells were observed as clumps whereas an RPE monolayer was visible in rats transplanted with the patch. These findings suggest that this approach may improve visual function at least in the short term in a few patients suffering from advanced stages of dry AMD. Currently, there is an ongoing clinical trial at phase I/IIa (NCT02590692) to study the safety of subretinal implantation of human embryonic stem cells seeded onto a parylene membrane [114].

The retinal scaffolds fabricated via photolithography are listed in Table 3.

Table 3. A summary of retinal scaffolds fabricated by lithography techniques. This table highlights the fabrication technique, the structure of scaffold and the cell type used. RPE: retinal pigment epithelium; ESC: embryonic stem cells.

Fabrication Technique	Scaffold Structure	Cell Type	Research Stage	Reference
Photolithography	Porous scaffold	Mouse retinal progenitor cells	In vitro	[110]
Photolithography	Thin film scaffold	Mouse retinal progenitor cells	In vitro	[111]
Photolithography	Porous scaffold	Mouse retinal progenitor cells	In vitro and in vivo (mouse)	[112]
Photolithography + wet and ion etching	Mesh-supported submicron membrane	Human cell line ARPE-19/H9-RPE	In vitro	[113]
Photolithography + wet and ion etching	Mesh-supported submicron membrane	Human ESC-RPE cells	In vitro and in vivo (rat)	[8]

3.4. 3D Bioprinting

To overcome the lack of a fine control of structural aspects, additive manufacturing (AM) techniques have been introduced into the tissue engineering area. AM is a process by which a digital model for a 3D object is assembled in a layer-by-layer manner [115]. AM advantages include the ability to fabricate complex geometries with multimaterial parts [115]. Among these techniques, 3D printing plays a key role in developing personalized treatments, surgical planning and the testing deployment of devices in realistic pathways due to its potential to fabricate patient-specific 3D structures in a cost-, time- and waste-effective manner [116]. In the medical field, the latest promising evolution of 3D printing is 3D bioprinting.

3D bioprinting is an innovative biofabrication strategy, which allows the precise positioning of non-living materials, as in 3D printing, and living materials in a prescribed 3D hierarchical organization [117]. During the bioprinting process, the bioinks, composed of cells embedded in biocompatible materials, are dispensed to form the functional desired structures. 3D bioprinting aims to create 3D bioengineered structures serving in regenerative medicine, pharmacokinetics and basic cell biology studies [117]. Such 3D bioprinted constructs can be then cultured in bioreactor systems to obtain mature functional tissues and organs [118]. A few examples of 3D bioprinted scaffolds are represented by synthetic skin to be transplanted onto patients with burn injuries, heart valve replication and bionic ears [119–121]. Currently, there are mainly three types of bioprinting systems (laser-based, inkjet-based and extrusion-based) characterized by a high deposition accuracy, stability and cell viability [117].

Laser-based technology exploits a pulsed laser source and an optical path to focus a laser on a target from which the bioink is printed and deposited onto a substrate. The target is composed of a glass slide, a metal slide and bioink. The laser is focused through the glass slide onto the metal slide inducing a vaporization of the metal-absorbing layer, thus resulting in the production of a jet of bioink. Many researchers have demonstrated that cells were highly viable after laser-based 3D bioprinting [122]. However, cell placement accuracy can be challenging and, in addition, this technology is expensive and relatively slow.

On the contrary, inkjet-based bioprinting is an inexpensive and simple to use technology while offering a relatively high resolution and cell viability [122]. In inkjet bioprinting, small droplets of bioink are ejected from a nozzle through microheater or piezoelectric systems and then dispensed onto a substrate [122]. Inkjet bioprinting was investigated to seed RGCs and glial cells. The technology did not affect the survival and the growth of rat RGCs and glial cells compared with cells seeded onto tissue culture plates [123]. This result opens the way for developing a printed construct to be used in retinal regeneration. Recently, a 3D in vitro retina model comprising of RPE cells and photoreceptors was fabricated using inkjet-based technology to study the interaction between the layers in AMD disease [124]. According to the authors, after bioprinting both cell types were correctly positioned in a layered structure and expressed specific proteins such as tight junction-associated protein ZO-1 in the RPE layer and light-sensitive proteins in the photoreceptor layer [124]. The use of inkjet-printed structures for tissue engineering is limited by their low mechanical properties and their long term durability. Moreover, due to the presence of the nozzle, clogging issues are common with viscous bioink [122].

The extrusion-based technology relies on the extrusion of continuous filaments of bioink through a nozzle using a driving force, i.e., a piston, a screwing system or pneumatic pressure. It is the most versatile bioprinting process as it enables the printing of the broadest range of bioink viscosities [122]. Additionally, it allows clinically relevant constructs to be obtained in terms of size and shape. However, the rheological requirements of the bioink are stringent [125]. Extrusion-based bioprinting was investigated to produce a 3D in vitro model of the RPE and photoreceptor layers [126]. The RPE cell line was bioprinted with a precise pattern and allowed to form a monolayer in 14 days followed by the bioprinting of the photoreceptor cell line [126]. Such a bioprinted construct could be meaningful for biomedical applications such as disease research and high-throughput screening.

So far, the bioprinted retinal construct has been used only as in vitro models that could be used in the future for basic research and drug screening (Table 4). However, this technology is very promising also for retinal tissue engineering due to the possibility of generating layer-by-layer a 3D complex multicellular stratified structure, which mimics the retina architecture and could be transplanted as potential therapies for retinal diseases involving the degeneration or dysfunction of multiple retinal layers.

Table 4. A summary of 3D bioprinting techniques used in retinal applications. This table highlights the fabrication technique, the structure of the scaffold and the cell type used. RGC: retinal ganglion cells.

Fabrication Technique	Scaffold Structure	Cell Type	Research Stage	Reference
Inkjet bioprinting	Not applicable	Rat RGCs and retinal glia	In vitro	[123]
Inkjet bioprinting	3D bilayer retina model	Human cell line ARPE-19 and pig photoreceptors	In vitro	[124]
Extrusion bioprinting	3D bilayer retina model	Human cell line ARPE-19 and Y79	In vitro	[126]

3.5. Hybrid Approach

Recently, researchers have focused on combining two or more fabrication approaches for maximal treatment efficacy (Table 5). Tan et al. combined SC/PL and TIPS to develop ultrathin polymer membranes as a prosthetic Bruch's membrane for an RPE replacement [127]. Solvent casting and photolithography were exploited to produce a polyester-based micropatterned film for RPE cells [128]. Shi et al. developed a hybrid approach to fabricate a 3D in vitro model that comprised an artificial Bruch's membrane, a bioprinted ARPE-19 cell monolayer and a Y79 cell-laden alginate/pluronic bioink [129]. This model could be used in the future to study the pathological mechanisms underlying AMD. Another hybrid approach based on electrospinning and 3D bioprinting has been established for the replacement of RGCs as a promising therapy for glaucoma [130]. RGCs were precisely seeded onto an electrospun scaffold via a thermal inkjet 3D cell printing technique. The electrospun scaffold guided the growth of the RGCs [130].

Table 5. A summary of hybrid approaches used in retinal applications. This table highlights the fabrication technique, the structure of the scaffold and the cell type used. RGC: retinal ganglion cells.

Fabrication Technique	Scaffold Structure	Cell Type	Research Stage	Reference
SC/PL + TIPS	Ultrathin, free-standing, porous membrane	Human cell line ARPE-19	In vitro	[127]
SC + photolithography	Micropatterned film	Human cell line D407	In vitro	[128]
SC + 3D bioprinting	3D construct: ultrathin membrane + bilayer model	Human cell line ARPE-19 and Y79	In vitro	[129]
Electrospinning + 3D bioprinting	Nanofibrous membrane + bioprinted cells	Rat RGCs	In vitro	[130]

4. Conclusions

Regenerative medicine research and optimization carries enormous hope as means to restore visual function compromised by retinal diseases. In this review, we reported both the conventional and recently developed methods to produce scaffolds for retinal tissue engineering. We believe that innovative tissue engineering techniques open the way for new therapies based on retinal grafts that could be implanted and reverse cell loss. Moreover, new tissue engineering strategies in conjunction with gene therapy holds great promise for the development of innovative clinical treatments to cure retinal diseases.

Funding: This research received no external funding.

Conflicts of Interest: The authors declare no conflict of interest.

References

1. Hoon, M.; Okawa, H.; Della Santina, L.; Wong, R.O.L. Functional architecture of the retina: Development and disease. *Prog. Retin. Eye Res.* **2014**, *42*, 44–84. [CrossRef] [PubMed]
2. Strauss, O. The retinal pigment epithelium in visual function. *Prog. Retin. Eye Res.* **2005**, *85*, 845–881. [CrossRef]
3. Adijanto, J.; Banzon, T.; Jalickee, S.; Wang, N.S.; Miller, S.S. CO_2-induced ion and fluid transport in human retinal pigment epithelium. *J. Gen. Physiol.* **2009**, *133*, 603–622. [CrossRef] [PubMed]
4. Kiser, P.D.; Golczak, M.; Palczewski, K. Chemistry of the Retinoid (Visual) Cycle. *Chem. Rev.* **2014**, *114*, 194–232. [CrossRef] [PubMed]
5. Murakami, Y.; Notomi, S.; Hisatomi, T.; Nakazawa, T.; Ishibashi, T.; Miller, J.W.; Vavvas, D.G. Photoreceptor cell death and rescue in retinal detachment and degenerations. *Prog. Retin. Eye Res.* **2014**, *37*, 1–55. [CrossRef]
6. Curcio, C.; Johnson, M. Structure, function, and pathology of Bruch's membrane. In *Retina*, 5th ed.; Elsevier: London, UK, 2013; Volume 1, pp. 466–481.
7. Booji, J.C.; Baas, D.C.; Beisekeeva, J.; Gorgels, T.G.M.F.; Bergen, A.A.B. The dynamic nature of Bruch's membrane. *Prog. Retin. Eye Res.* **2010**, *29*, 1–18. [CrossRef]
8. Diniz, B.; Thomas, P.; Thomas, B.; Ribeiro, R.; Hu, Y.; Brant, R.; Ahuja, A.; Zhu, D.; Liu, L.; Koss, M.; et al. Subretinal implantation of retinal pigment epithelial cells derived from human embryonic stem cells: Improved survival when implanted as monolayer. *Investig. Ophthalmol. Vis. Sci.* **2013**, *26*, 5087–5096. [CrossRef]
9. Dutt, K.; Cao, Y. Engineering retina from human retinal progenitors (cell lines). *Tissue Eng. A* **2009**, *15*, 1401–1413. [CrossRef]

10. Wong, W.L.; Su, X.; Li, X.; Cheung, C.M.; Klein, R.; Cheng, C.Y.; Wong, T.Y. Global prevalence of age-related macular degeneration and disease burden projection for 2020 and 2040: A systematic review and meta-analysis. *Lancet Glob. Health* **2014**, *2*, 106–116. [CrossRef]
11. Bird, A.C. Pathogenic mechanisms in age-related macular degeneration. In *Retina*, 5th ed.; Elsevier Inc.: London, UK, 2013; Volume 1, pp. 1145–1149.
12. Fernandez-Robredo, P.; Sancho, A.; Johnen, S.; Recalde, S.; Gama, N.; Thumann, G.; Groll, J.; Garcia-Layana, A. Current Treatment Limitations in Age-Related Macular Degeneration and Future Approaches Based on Cell Therapy and Tissue Engineering. *Ophthalmology* **2014**, 1–13. [CrossRef]
13. Jonas, J.B.; Aung, T.; Bourne, R.R.; Bron, A.M.; Ritch, R.; Panda-Jonas, S. Glaucoma. *Lancet* **2017**, *390*, 2183–2193. [CrossRef]
14. Weinreb, R.N.; Aung, T.; Medeiros, F.A. The Pathophysiology and Treatment of Glaucoma. *JAMA* **2015**, *311*, 1901–1911. [CrossRef] [PubMed]
15. Sahle, F.F.; Kim, S.; Niloy, K.K.; Tahia, F.; Fili, C.V.; Cooper, E.; Hamilton, D.J.; Lowe, T.L. Nanotechnology in Regenerative Ophthalmology. *Adv. Drug Deliv. Rev.* **2019**, *148*, 290–307. [CrossRef] [PubMed]
16. Heier, J.S.; Kherani, S.; Desai, S.; Dugel, P.; Kaushal, S.; Cheng, S.H.; Delacono, C.; Purvis, A.; Richards, S.; Le-Halpere, A.; et al. Intravitreous injection of AAV2-sFLT01 in patients with advanced neovascular age-related macular degeneration: A phase 1, open-label trial. *Lancet* **2017**, *390*, 50–61. [CrossRef]
17. Georgiou, M.; Fujinami, K.; Michaelides, M. Inherited retinal diseases: Therapeutics, clinical trials and end points—A review. *Clin. Exp. Ophthalmol.* **2021**, *49*, 270–288. [CrossRef]
18. Tan, Y.S.E.; Shi, P.J.; Choo, C.J.; Laude, A.; Yeong, W.Y. Tissue engineering of retina and Bruch's membrane: A review of cells, materials and processes. *Br. J. Ophthalmol.* **2018**, *102*, 1182–1187. [CrossRef]
19. Fu, L.; Kwok, S.S.; Chan, Y.K.; Lai, J.S.M.; Pan, W.; Nie, L.; Shih, K.C. Therapeutic Strategies for Attenuation of Retinal Ganglion Cell Injury in Optic Neuropathies: Concepts in Translational Research and Therapeutic Implications. *Biomed. Res. Int.* **2019**, *2019*, 1–10. [CrossRef]
20. McGill, T.J.; Osborne, L.; Lu, B.; Stoddard, J.; Huhn, S.; Tsukamoto, A.; Capela, A. Subretinal Transplantation of Human Central Nervous System Stem Cells Stimulates Controlled Proliferation of Endogenous Retinal Pigment Epithelium. *Transl. Vis. Sci. Technol.* **2019**, *8*, 43–54. [CrossRef]
21. Lu, B.; Malcuit, C.; Wang, S.; Girman, S.; Francis, P.; Lemieux, L.; Lanza, R.; Lund, R. Long-term safety and function of RPE from human embryonic stem cells in preclinical models of macular degeneration. *Stem Cells* **2009**, *27*, 2126–2135. [CrossRef] [PubMed]
22. Singh, M.S.; Park, S.S.; Albini, T.A.; Canto-Soler, M.V.; Klassen, H.; MacLaren, R.E.; Takahashi, M.; Nagiel, A.; Schwartz, S.D.; Bharti, K. Retinal stem cell transplantation: Balancing safety and potential. *Prog. Retin. Eye Res.* **2020**, *75*, 100779. [CrossRef] [PubMed]
23. Mandai, M.; Watanabe, A.; Kurimoto, Y. Autologous Induced Stem-Cell-Derived Retinal Cells for Macular Degeneration. *N. Engl. J. Med.* **2017**, *376*, 1038–1046. [CrossRef] [PubMed]
24. Kubota, A.; Nishida, K.; Yamato, M.; Yang, J.; Kikuchi, A.; Okano, T.; Tano, Y. Transplantable retinal pigment epithelial cell sheets for tissue engineering. *Biomaterials* **2006**, *27*, 3639–3644. [CrossRef]
25. Chen, G.; Qi, Y.; Niu, L.; Dl, T.; Zhong, J.; Fang, T.; Yan, W. Application of the cell sheet technique in tissue engineering. *Biomed. Rep.* **2015**, *3*, 749–757. [CrossRef]
26. Haraguchi, Y.; Shimizu, T.; Yamato, M.; Okano, T. Scaffold-free tissue engineering using cell sheet technology. *RSC Adv.* **2012**, *2*, 2184–2190. [CrossRef]
27. Langer, R.; Vacanti, J. Advances in Tissue Engineering. *J. Pediatr. Surg.* **2016**, *51*, 8–12. [CrossRef]
28. Asnaghi, M.A.; Candiani, G.; Farè, S.; Fiore, G.B.; Petrini, P.; Raimondi, M.T.; Soncini, M.; Mantero, S. Trends in biomedical engineering: Focus on Regenerative Medicine. *J. Appl. Biomater. Biomech.* **2011**, *9*, 73–86. [CrossRef]
29. Curtis, M.W.; Russel, B. Cardiac Tissue Engineering. *J. Cardiovasc. Nurs.* **2009**, *24*, 87–92. [CrossRef]
30. Shafiee, A.; Atala, A. Tissue Engineering: Toward a New Era of Medicine. *Annu. Rev. Med.* **2017**, *14*, 29–40. [CrossRef]
31. Del Priore, L.V.; Tezel, T.H. Reattachment rate of human retinal pigment epithelium to layers of human Bruch's membrane. *Arch. Ophthalmol.* **1998**, *116*, 335–341. [CrossRef] [PubMed]
32. Tezel, T.H.; Kaplan, H.J.; Del Priore, L.V. Fate of human retinal pigment epithelial cells seeded onto layers of human Bruch's membrane. *Investig. Ophthalmol. Vis. Sci.* **1999**, *40*, 467–476.
33. Tezel, T.H.; Del Priore, L.V. Repopulation of different layers of host human Bruch's membrane by retinal pigment epithelial cell grafts. *Investig. Ophthalmol. Vis. Sci.* **1999**, *40*, 767–774.
34. Gullapalli, V.K.; Sugino, I.K.; Van Patten, Y.; Shah, S.; Zarbin, M.A. Impaired RPE survival on aged submacular humanBruch's membrane. *Exp. Eye Res.* **2005**, *80*, 235–248. [CrossRef]
35. Gullapalli, V.K.; Sugino, I.K.; Van Patten, Y.; Shah, S.; Zarbin, M.A. Retinal pigment epithelium resurfacing of aged submacular human Bruch's membrane. *Trans. Am. Ophthalmol. Soc.* **2004**, *102*, 123–137.
36. Tezel, T.H.; Del Priore, L.V.; Kaplan, H.J. Reengineering of aged Bruch's membrane to enhance retinal pigment epithelium repopulation. *Investig. Ophthalmol. Vis. Sci.* **2004**, *45*, 3337–3348. [CrossRef]
37. Del Priore, L.V.; Geng, L.; Tezel, T.H.; Kaplan, H.J. Extracellular matrix ligands promote RPE attachment to inner Bruch's membrane. *Curr. Eye Res.* **2002**, *25*, 79–89. [CrossRef] [PubMed]

38. Sugino, I.K.; Gullapalli, V.K.; Sun, Q.; Wang, J.; Nunes, C.F.; Cheewatrakoolpong, N.; Johnson, A.C.; Degner, B.C.; Hua, J.; Liu, T.; et al. Cell-deposited matrix improves retinal pigment epithelium survival on aged submacular human Bruch's membrane. *Investig. Ophthalmol. Vis. Sci.* **2011**, *52*, 1345–1358. [CrossRef]
39. Sugino, I.K.; Rapista, A.; Sun, Q.; Wang, J.; Nunes, C.F.; Cheewatrakoolpong, N.; Zarbin, M.A. A method to enhance cell survival on Bruch's membrane in eyes affected by age and age-related macular degeneration. *Investig. Ophthalmol. Vis. Sci.* **2011**, *52*, 9598–9609. [CrossRef] [PubMed]
40. Chaudhari, A.A.; Vig, K.; Baganizi, D.R.; Sahu, R.; Dixit, S.; Dennis, V.; Singh, S.R.; Pillai, S.R. Future Prospects for Scaffolding Methods and Biomaterials in Skin Tissue Engineering: A Review. *Int. J. Mol. Sci.* **2016**, *17*, 1974. [CrossRef] [PubMed]
41. Lam Van Ba, O.; Aharony, S.; Loutochin, O.; Corcos, J. Bladder tissue engineering: A literature review. *Adv. Drug Deliv. Rev.* **2015**, *82–83*, 31–37. [CrossRef]
42. Kessler, M.W.; Grande, D.A. Tissue engineering and cartilage. *Organogenesis* **2008**, *4*, 28–32. [CrossRef]
43. Wubneh, A.; Tsekoura, E.K.; Ayranci, C.; Uludag, H. Current state of fabrication technologies and materials for bone tissue engineering. *Acta Biomater.* **2018**, *80*, 1–30. [CrossRef]
44. Mertens, J.P.; Sugg, K.B.; Lee, J.D.; Larkin, L.M. Engineering muscle constructs for the creation of functional engineered musculoskeletal tissue. *Regen. Med.* **2014**, *9*, 89–100. [CrossRef] [PubMed]
45. Chirila, T.; Barnard, Z.; Zainuddin; Harkin, D.G.; Schwab, I.R.; Hirst, L. Bombyx mori silk fibroin membranes as potential substrata for epithelial constructs used in the management of ocular surface disorders. *Tissue Eng. Part A* **2008**, *14*, 1203–1211. [CrossRef] [PubMed]
46. Siemann, U. Solvent cast technology—A versatile tool for thin film production. In *Scattering Methods and Properties of Polymer Materials: Progress in Colloid and Polymer Science*; Stribeck, N., Smarsly, B., Eds.; Springer: Berlin, Germany, 2005; Volume 130, pp. 1–14.
47. Giordano, G.G.; Thomson, R.C.; Ishaug, S.L.; Mikos, A.G.; Cumber, S.; Garcia, C.A.; Lahiri-Munir, D. Retinal pigment epithelium cells cultured on synthetic biodegradable polymers. *J. Biomed. Res.* **1997**, *34*, 87–93. [CrossRef]
48. Lai, J.; Li, Y. Evaluation of cross-linked gelatin membranes as delivery carriers for retinal sheets. *Mater. Sci. Eng. C* **2010**, *30*, 677–685. [CrossRef]
49. Lai, J. Influence of solvent composition on the performance of carbodiimide cross-linked gelatin carriers for retinal sheet delivery. *J. Mater. Sci. Mater. Med.* **2013**, *24*, 2201–2210. [CrossRef]
50. Shadforth, A.M.; Suzuki, S.; Alzonne, R.; Edwards, G.A.; Richardson, N.A.; Chirila, T.V.; Harkin, D.G. Incorporation of human recombinant tropoelastin into silk fibroin membranes with the view of repairing Bruch's membrane. *J. Funct. Biomater.* **2015**, *6*, 946–962. [CrossRef]
51. Singh, A.K.; Srivastava, G.K.; Martin, L.; Alonso, M.; Pastor, J.C. Bioactive substrates for human retinal pigment epithelial cell growth from elastin-like recombinamers. *J. Biomed. Mater. Res. A* **2014**, *102*, 639–646. [CrossRef]
52. Galloway, C.A.; Dalvi, S.; Shadforth, A.M.A.; Suzuki, S.; Wilson, M.; Kuai, D.; Hashim, A.; MacDonald, L.A.; Gamm, D.M.; Harkin, D.G.; et al. Characterization of Human iPSC-RPE on a Prosthetic Bruch's Membrane Manufactured from Silk Fibroin. *Investig. Ophthalmol. Vis. Sci.* **2018**, *59*, 2792–2800. [CrossRef] [PubMed]
53. Tezcaner, A.; Bugra, K.; Hasirici, V. Retinal pigment epithelium cell culture on surface modified poly(hydroxybutyrate-co-hydroxyvalerate) thin films. *Biomaterials* **2003**, *24*, 4573–4583. [CrossRef]
54. Prasad, A.; Sankar, M.R.; Katiyar, V. State of Art on Solvent Casting Particulate Leaching Method for Orthopedic Scaffolds Fabrication. *Mater. Today* **2017**, *4*, 898–907. [CrossRef]
55. Subia, B.; Kundu, J.; Kundu, S.C. Biomaterial scaffold fabrication techniques for potential tissue engineering applications. In *Tissue Engineering*; Eberli, D., Ed.; InTech: Rijeka, Croatia, 2010; pp. 141–157.
56. Shadforth, A.M.; George, K.A.; Kwan, A.S.; Chirila, T.V.; Harkin, D.G. The cultivation of human retinal pigment epithelium on Bombyx mori silk fibroin. *Biomaterials* **2012**, *33*, 4110–4117. [CrossRef]
57. McHugh, K.J.; Tao, S.L.; Saint-Geniez, M. Porous poly(epsilon-caprolactone) scaffolds for retinal pigment epithelium transplantation. *Investig. Opthalmol. Vis. Sci.* **2014**, *55*, 1754–1762. [CrossRef]
58. Calejo, M.T.; Ilmarinen, T.; Vuorimaa-Laukkanen, E.; Talvitie, E.; Hakola, H.M.; Skottman, H.; Kellomaki, M. Langmuir-Schaefer film deposition onto honeycomb porous films for retinal tissue engineering. *Acta Biomater.* **2017**, *54*, 138–149. [CrossRef]
59. Worthington, K.S.; Wiley, L.A.; Guymon, C.A.; Salem, A.K.; Tucker, B.A. Differentiation of Induced Pluripotent Stem Cells to Neural Retinal Precursor Cells on Porous Poly-Lactic-co-Glycolic Acid Scaffolds. *J. Ocul. Pharmacol. Ther.* **2016**, *32*, 310–316. [CrossRef]
60. Janik, H.; Marzec, M. A review: Fabrication of porous polyurethane scaffolds. *Mater. Sci. Eng. C Mater. Biol. Appl.* **2015**, *48*, 586–591. [CrossRef] [PubMed]
61. El-Sherbiny, I.M.; Yacoub, M. Hydrogel scaffolds for tissue engineering: Progress and challenges. *Glob. Cardiol. Sci. Pract.* **2013**, 3, 316–342. [CrossRef]
62. Maitra, J.; Shukala, V.K. Cross-linking in Hydrogels—A Review. *Am. J. Polym. Sci* **2014**, *2*, 25–31. [CrossRef]
63. Park, J.H.; Shin, E.Y.; Shin, M.E.; Choi, M.J.; Carlomagno, C.; Song, J.E.; Khang, G. Enhanced retinal pigment epithelium (RPE) regeneration using curcumin/alginate hydrogels: In vitro evaluation. *Int. J. Biol. Macromol.* **2018**, *117*, 546–552. [CrossRef] [PubMed]

64. Kim, H.S.; Kim, D.; Jeong, Y.W.; Choi, M.J.; Lee, G.W.; Thangavelu, M.; Song, J.E.; Khang, G. Engineering retinal pigment epithelial cells regeneration for transplantation in regenerative medicine using PEG/Gellan gum hydrogels. *Int. J. Biol. Macromol.* **2019**, *130*, 220–228. [CrossRef] [PubMed]
65. Hunt, N.C.; Hallam, D.; Karimi, A.; Mellough, C.B.; Chen, J.; Steel, D.H.W.; Lako, M. 3D culture of human pluripotent stem cells in RGD-alginate hydrogel improves retinal tissue development. *Acta Biomater.* **2017**, *49*, 329–343. [CrossRef] [PubMed]
66. Gandhi, J.K.; Manzar, Z.; Bachman, L.A.; Andrews-Pfannkoch, C.; Knudsen, T.; Hill, M.; Schmidt, H.; Iezzi, R.; Pulido, J.S.; Marmorstein, A.D. Fibrin hydrogels as a xenofree and rapidly degradable support for transplantation of retinal pigment epithelium monolayers. *Acta Biomater.* **2018**, *67*, 134–146. [CrossRef] [PubMed]
67. Lu, J.T.; Lee, C.J.; Bent, S.F.; Fishman, H.A.; Sabelman, E.E. Thin collagen film scaffolds for retinal epithelial cell culture. *Biomaterials* **2007**, *28*, 1486–1494. [CrossRef] [PubMed]
68. Hertz, J.; Robinson, R.; Valenzuela, D.A.; Lavik, E.B.; Goldberg, J.L. A tunable synthetic hydrogel system for culture of retinal ganglion cells and amacrine cells. *Acta Biomater.* **2013**, *9*, 7622–7629. [CrossRef] [PubMed]
69. Roozafzoon, R.; Lashay, A.; Vasei, M.; Ai, J.; Khoshzaban, A.; Keshel, S.H.; Barabadi, Z.; Bahrami, H. Dental pulp stem cells differentiation into retinal ganglion-like cells in a three dimensional network. *Biochem. Biophys. Res. Commun.* **2015**, *457*, 154–160. [CrossRef]
70. Soleimannejad, M.; Ebrahimi-Barough, S.; Soleimani, M.; Nadri, S.; Tavangar, S.M.; Roohipoor, R.; Yazdankhah, M.; Bayat, N.; Riazi-Esfahani, M.; Ai, J. Fibrin gel as a scaffold for photoreceptor cells differentiation from conjunctiva mesenchymal stem cells in retina tissue engineering. *Artif. Cells Nanomed. Biotechnol.* **2018**, *46*, 805–814. [CrossRef]
71. Conoscenti, G.; Carrubba, V.L.; Brucato, V. A versatile technique to produce porous polymeric scaffolds: The Thermally Induced Phase Separation (TIPS) method. *Arch. Chem. Res.* **2017**, *1*. [CrossRef]
72. Lu, T.; Li, Y.; Chen, T. Techniques for fabrication and construction of three-dimensional scaffolds for tissue engineering. *Int. J. Nanomed.* **2013**, *8*, 337–350. [CrossRef]
73. Akbarzadeh, R.; Yousefi, A. Effects of processing parameter in thermally induced phase separation technique on porous architecture of scaffolds for bone tissue engineering. *J. Biomed. Mater. Res. B Appl. Biomater.* **2014**, *102*, 1304–1315. [CrossRef]
74. Turnbull, G.; Clarke, J.; Picard, F.; Riches, P.; Jia, L.; Han, F.; Li, B.; Shu, W. 3D bioactive composite scaffolds for bone tissue engineering. *Bioact. Mater.* **2018**, *3*, 278–314. [CrossRef]
75. Martinez-Perez, C.A.; Olivas-Armendariz, I.; Castro-Carmona, J.S.; Garcia-Casillas, P.E. Scaffolds for tissue engineering via thermally induced phase separation. In *Advances in Regenerative Medicine*; Wislet-Gendebien, S., Ed.; InTech Open: London, UK, 2011; pp. 275–294.
76. Thomson, H.A.J.; Treharne, A.J.; Walker, P.; Grossel, M.C.; Lotery, A.J. Optimisation of polymer scaffolds for retinal pigment epithelium (RPE) cell transplantation. *Br. J. Ophthalmol.* **2011**, *95*, 563–568. [CrossRef]
77. Lavik, E.B.; Klassen, H.; Warfvinge, K.; Langer, R.; Young, M.J. Fabrication of degradable polymer scaffolds to direct the integration and differentiation of retinal progenitors. *Biomaterials* **2005**, *26*, 3187–3196. [CrossRef] [PubMed]
78. Hotaling, N.A.; Khristov, V.; Wan, Q.; Sharma, R.; Jha, B.S.; Lotfi, M.; Maminishkis, A.; Simon, C.G., Jr.; Bharti, K. Nanofiber Scaffold-Based Tissue-Engineered Retinal Pigment Epithelium to Treat Degenerative Eye Diseases. *J. Ocul. Pharmacol. Ther.* **2016**, *32*, 272–285. [CrossRef] [PubMed]
79. Rahmati, M.; Mills, D.K.; Urbanska, A.M.; Reza Saeb, M.; Venugopal, J.R.; Ramakrishna, S.; Mozafari, M. Electrospinning for Tissue Engineering Applications. *Prog. Mater. Sci.* **2020**, 100721. [CrossRef]
80. Bhardwaj, N.; Kundu, S.C. Electrospinning: A fascinating fiber fabrication technique. *Biotechnol. Adv.* **2010**, *28*, 325–347. [CrossRef]
81. Beachley, V.; Wen, X. Effect of electrospinning parameters on the nanofiber diameter and length. *Mater. Sci. Eng. C Mater. Biol. Appl.* **2009**, *29*, 663–668. [CrossRef] [PubMed]
82. Hasan, A.; Memic, A.; Annabi, N.; Hossain, M.; Paul, A.; Dokmeci, M.R.; Dehghani, F.; Khademhosseini, A. Electrospun scaffolds for tissue engineering of vascular grafts. *Acta Biomater.* **2014**, *10*, 11–25. [CrossRef]
83. Kitsara, M.; Agbulut, O.; Kontziampasis, D.; Chen, Y.; Menasché, P. Fibers for heart: A critical review on electrospinning for cardiac tissue engineering. *Acta Biomater.* **2017**, *48*, 20–40. [CrossRef]
84. Xie, J.; MacEwan, M.R.; Schwartz, A.G.; Xia, Y. Electrospun nanofibers for neural tissue engineering. *Nanoscale* **2010**, *2*, 35–44. [CrossRef]
85. Warnke, P.H.; Alamein, M.; Skabo, S.; Stephens, S.; Bourke, R.; Heiner, P.; Liu, Q. Primordium of an artificial Bruch's membrane made of nanofibers for engineering of retinal pigment epithelium cell monolayers. *Acta Biomater.* **2013**, *9*, 9414–9422. [CrossRef] [PubMed]
86. Xiang, P.; Wu, K.C.; Zhu, Y.; Xiang, L.; Li, C.; Chen, D.L.; Chen, F.; Xu, G.; Wang, A.; Li, M.; et al. A novel Bruch's membrane mimetic electrospun substrate scaffold for human retinal pigment epithelium cells. *Biomaterials* **2014**, *35*, 9777–9788. [CrossRef] [PubMed]
87. Zhang, D.; Ni, N.; Chen, J.; Yao, Q.; Shen, B.; Zhang, Y.; Zhu, M.; Wang, Z.; Ruan, J.; Wang, J.; et al. Electrospun SF/PLCL nanofibrous membrane: A potential scaffold for retinal progenitor cell proliferation and differentiation. *Sci. Rep.* **2015**, *5*, 14326. [CrossRef]

88. Popelka, Š.; Studenovská, H.; Abelová, L.; Ardan, T.; Studený, P.; Straňák, Z.; Klíma, J.; Dvořánková, B.; Kotek, J.; Hodan, J.; et al. A frame-supported ultrathin electrospun polymer membrane for transplantation of retinal pigment epithelial cells. *Biomed. Mater.* **2015**, *10*. [CrossRef] [PubMed]
89. Surrao, D.C.; Greferath, U.; Chau, Y.Q.; Skabo, S.J.; Huynh, M.; Shelat, K.J.; Limnios, I.J.; Fletcher, E.L.; Liu, Q. Design, development and characterization of synthetic Bruch's membranes. *Acta Biomater.* **2017**, *64*, 357–376. [CrossRef] [PubMed]
90. Thieltges, F.; Stanzel, B.V.; Liu, Z.; Holz, F.G. A nanofibrillar surface promotes superior growth characteristics in cultured human retinal pigment epithelium. *Ophthalmic Res.* **2011**, *46*, 133–140. [CrossRef]
91. Noorani, B.; Tabandeh, F.; Yazdian, F.; Soheili, Z.; Shakibaie, M.; Rahmani, S. Thin natural gelatin/chitosan nanofibrous scaffolds for retinal pigment epithelium cells. *Int. J. Polym. Mater. Polym. Biomater.* **2017**, *67*, 754–763. [CrossRef]
92. Liu, Z.; Yu, N.; Holz, F.G.; Yang, F.; Stanzel, B.V. Enhancement of retinal pigment epithelial culture characteristics and subretinal space tolerance of scaffolds with 200 nm fiber topography. *Biomaterials* **2014**, *35*, 2837–2850. [CrossRef]
93. Shahmoradi, S.; Yazdian, F.; Tabandeh, F.; Soheili, Z.S.; Hatamian Zarami, A.S.; Navaei-Nigjeh, M. Controlled surface morphology and hydrophilicity of polycaprolactone toward human retinal pigment epithelium cells. *Mater. Sci. Eng. C Mater. Biol. Appl.* **2017**, *1*, 300–309. [CrossRef]
94. Sorkio, A.; Porter, P.J.; Juuti-Uusitalo, K.; Meenan, B.J.; Skottman, H.; Burke, G.A. Surface Modified Biodegradable Electrospun Membranes as a Carrier for Human Embryonic Stem Cell-Derived Retinal Pigment Epithelial Cells. *Tissue Eng. Part A* **2015**, *21*, 2301–2314. [CrossRef]
95. Sharma, R.; Khristov, V.; Rising, A.; Jha, B.S.; Dejene, R.; Hotaling, N.; Li, Y.; Stoddard, J.; Stankewicz, C.; Wan, Q.; et al. Clinical-grade stem cell-derived retinal pigment epithelium patch rescues retinal degeneration in rodents and pigs. *Sci. Transl. Med.* **2019**, *16*, 475. [CrossRef]
96. Tian, Y.; Zonca, M.R.; Imbrogno, J.; Unser, A.M.; Sfakis, L.; Temple, S.; Belfort, G.; Xie, Y. Polarized, Cobblestone, Human Retinal Pigment Epithelial Cell Maturation on a Synthetic PEG Matrix. *ACS Biomater. Sci. Eng.* **2017**, *3*, 890–902. [CrossRef] [PubMed]
97. Rahmani, S.; Tabandeh, F.; Faghihi, S.; Amoabediny, G.; Shakibaie, M. Fabrication of poly(ε-caprolactone)/gelatin nanofibrous scaffolds for retinal tissue engineering. *Int. J. Polym. Mater. Polym. Biomater.* **2018**, *67*, 27–35. [CrossRef]
98. Da Silva, G.R.; Lima, T.H.; Oréfice, R.L.; Fernandes-Cunha, G.M.; Silva-Cunha, A.; Zhao, M.; Behar-Cohen, F. In vitro and in vivo ocular biocompatibility of electrospun poly(ε-caprolactone) nanofibers. *Eur. J. Pharm. Sci.* **2015**, *20*, 9–19. [CrossRef] [PubMed]
99. Chen, H.; Fan, X.; Xia, J.; Chen, P.; Zhou, X.; Huang, J.; Yu, J.; Gu, P. Electrospun chitosan-graft-poly (ε-caprolactone)/poly (ε-caprolactone) nanofibrous scaffolds for retinal tissue engineering. *Int. J. Nanomed.* **2011**, *6*, 453–461. [CrossRef]
100. Belgio, B.; Boschetti, F.; Mantero, S. Towards an In Vitro Retinal Model to Study and Develop New Therapies for Age-Related Macular Degeneration. *Bioengineering* **2021**, *8*, 18. [CrossRef]
101. Kador, K.E.; Montero, R.B.; Venugopalan, P.; Hertz, J.; Zindell, A.N.; Valenzuela, D.A.; Uddin, M.S.; Lavik, E.B.; Muller, K.J.; Andreopoulos, F.M.; et al. Tissue engineering the retinal ganglion cell nerve fiber layer. *Biomaterials* **2013**, *34*, 4242–4250. [CrossRef]
102. Kador, K.E.; Alsehli, H.S.; Zindell, A.N.; Lau, L.W.; Andreopoulos, F.M.; Watson, B.D.; Goldberg, J.L. Retinal ganglion cell polarization using immobilized guidance cues on a tissue-engineered scaffold. *Acta Biomater.* **2014**, *10*, 4939–4946. [CrossRef]
103. Li, K.; Zhong, X.; Yang, S.; Luo, Z.; Li, K.; Liu, Y.; Cai, S.; Gu, H.; Lu, S.; Zhang, H.; et al. HiPSC-derived retinal ganglion cells grow dendritic arbors and functional axons on a tissue-engineered scaffold. *Acta Biomater.* **2017**, *54*, 117–127. [CrossRef]
104. Nadri, S.; Kazemi, B.; Eslaminejad, M.B.; Yazdani, S.; Soleimani, M. High yield of cells committed to the photoreceptor-like cells from conjunctiva mesenchymal stem cells on nanofibrous scaffolds. *Mol. Biol. Rep.* **2013**, *40*, 3883–3890. [CrossRef]
105. Dalton, P.D. Melt electrowriting with additive manufacturing principles. *Curr. Opin. Biomed. Eng.* **2017**, *2*, 49–57. [CrossRef]
106. Saidy, N.T.; Shabab, T.; Bas, O.; Rojas-Gonzalez, D.M.; Menne, M.; Henry, T.; Hutmacher, D.W.; Mela, P.; De-Juan-Pardo, E.M. Melt Electrowriting of Complex 3D Anatomically Relevant Scaffolds. *Front. Bioeng. Biotechnol.* **2020**. [CrossRef] [PubMed]
107. Brennan, C.M.; Eichholz, K.F.; Hoey, D.A. The effect of pore size within fibrous scaffolds fabricated using melt electrowriting on human bone marrow stem cell osteogenesis. *Biomed. Mater.* **2019**, *14*. [CrossRef] [PubMed]
108. Hewitt, E.; Mros, S.; McConnell, M.; Cabral, J.D.; Ali, A. Melt-electrowriting with novel milk protein/PCL biomaterials for skin regeneration. *Biomed. Mater.* **2019**, *14*, 055013. [CrossRef] [PubMed]
109. Tran, K.T.M.; Nguyen, T.D. Lithography-based methods to manufacture biomaterials at small scales. *J. Sci. Adv. Mater. Devices* **2017**, *2*, 1–14. [CrossRef]
110. Neeley, W.L.; Redenti, S.; Klassen, H.; Tao, S.; Desai, T.; Young, M.J.; Langer, R. A microfabricated scaffold for retinal progenitor cell grafting. *Biomaterials* **2008**, *29*, 418–426. [CrossRef] [PubMed]
111. Steedman, M.R.; Tao, S.L.; Klassen, H.; Desai, T.A. Enhanced differentiation of retinal progenitor cells using microfabricated topographical cues. *Biomed. Microdevices* **2010**, *12*, 363–369. [CrossRef]
112. Redenti, S.; Neeley, W.L.; Rompani, S.; Saigal, S.; Yang, J.; Klassen, H.; Langer, R.; Young, M.J. Engineering retinal progenitor cell and scrollable poly(glycerol-sebacate) composites for expansion and subretinal transplantation. *Biomaterials* **2009**, *30*, 3405–3414. [CrossRef]
113. Lu, B.; Zhu, D.; Hinton, D.; Humayun, M.S.; Tai, Y.C. Mesh-supported submicron parylene-C membranes for culturing retinal pigment epithelial cells. *Biomed. Microdevices* **2012**, *14*, 659–667. [CrossRef]
114. Kashani, A.H.; Lebkowski, J.S.; Rahhal, F.M.; Avery, R.L.; Salehi-Had, H.; Dang, W.; Lin, C.M.; Mitra, D.; Zhu, D.; Thomas, B.B.; et al. A bioengineered retinal pigment epithelial monolayer for advanced, dry age-related macular degeneration. *Sci. Transl. Med.* **2018**, *4*, 435. [CrossRef]

115. Gleadall, A.; Visscher, D.; Yang, J.; Thomas, D.; Segal, J. Review of additive manufactured tissue engineering scaffolds: Relationship between geometry and performance. *Burn. Trauma* **2018**, *6*, 19. [CrossRef] [PubMed]
116. Aimar, A.; Palermo, A.; Innocenti, B. The Role of 3D Printing in Medical Applications: A State of the Art. *J. Healthc. Eng.* **2019**, *10*. [CrossRef] [PubMed]
117. Moroni, L.; Burdick, J.A.; Highley, C.; Lee, S.J.; Morimoto, Y.; Takeuchi, S.; Yoo, J.J. Biofabrication strategies for 3D in vitro models and regenerative medicine. *Nat. Rev. Mater.* **2018**, *3*, 21–37. [CrossRef] [PubMed]
118. Mantero, S.; Sadr, N.; Riboldi, S.A.; Lorenzoni, S.; Montevecchi, F.M. A new electro-mechanical bioreactor for soft tissue engineering. *J. Appl. Biomater. Biomech.* **2007**, *5*, 107–116. [PubMed]
119. He, P.; Zhao, J.; Zhang, J.; Li, B.; Gou, Z.; Gou, M.; Li, X. Bioprinting of skin constructs for wound healing. *Burn. Trauma* **2018**, *6*. [CrossRef]
120. Vukievic, M.; Mosadegh, B.; Little, J.K.; Little, S.H. Cardiac 3D printing and its future directions. *JACC Cardiovasc. Imaging* **2017**, *10*, 171–184. [CrossRef]
121. Mannoor, M.S.; Jiang, Z.; James, T.; Kong, Y.L.; Malatesta, K.A.; Soboyejo, W.O.; Verma, N.; Gracias, D.H.; McAlpine, M.C. 3D printed bionic ears. *Nano Lett.* **2013**, *13*, 2634–2639. [CrossRef] [PubMed]
122. Li, J.; Chen, M.; Fan, X.; Zhou, H. Recent advances in bioprinting techniques: Approaches, applications and future prospects. *J. Transl. Med.* **2016**, *14*. [CrossRef]
123. Lorber, B.; Hsiao, W.K.; Hutchings, I.M.; Martin, K.R. Adult rat retinal ganglion cells and glia can be printed by piezoelectric inkjet printing. *Biofabrication* **2014**, *6*. [CrossRef]
124. Masaeli, E.; Forster, V.; Picaud, S.; Karamali, F.; Nasr-Esfahani, M.H.; Marquette, C. Tissue engineering of retina through high resolution 3-dimensional inkjet bioprinting. *Biofabrication* **2020**, *12*. [CrossRef]
125. Schwab, A.; Levato, R.; D'Este, M.; Piluso, S.; Eglin, D.; Malda, J. Printability and Sahe Fidelity of Bioiks in 3D Bioprinting. *Chem. Rev.* **2020**, *120*, 11028–11055. [CrossRef]
126. Shi, P.; Tan, Y.S.E.; Yeong, W.Y.; Li, H.Y.; Laude, A. A bilayer photoreceptor-retinal tissue model with gradient cell density design: A study of microvalve-based bioprinting. *J. Tissue Eng. Regen. Med.* **2018**, *12*, 1297–1306. [CrossRef] [PubMed]
127. Tan, E.Y.S.; Agarwala, S.; Yap, Y.L.; Tan, C.S.H.; Laude, A.; Yeong, W.Y. Novel method for the fabrication of ultrathin, free-standing and porous polymer membranes for retinal tissue engineering. *J. Mater. Chem. B* **2017**, *5*, 5616–5622. [CrossRef]
128. Zorlutuna, P.; Builles, N.; Damour, O.; Elsheikh, A.; Hasirci, V. Influence of keratocytes and retinal pigment epithelial cells on the mechanical properties of polyester-based tissue engineering micropatterned films. *Biomaterials* **2007**, *28*, 3489–3496. [CrossRef]
129. Shi, P.; Edgar, T.Y.S.; Yeong, W.Y.; Laude, A. Hybrid three-dimensional (3D) bioprinting of retina equivalent for ocular research. *Int. J. Bioprint.* **2017**, *3*. [CrossRef]
130. Kador, K.E.; Grogan, S.P.; Dorthé, E.W.; Venugopalan, P.; Malek, M.F.; Goldberg, J.L.; D'lima, D.D. Control of Retinal Ganglion Cell Positioning and Neurite Growth: Combining 3D Printing with Radial Electrospun Scaffolds. *Tissue Eng. Part A* **2016**, *22*, 286–294. [CrossRef] [PubMed]

Article

Alginate-Chitosan Microencapsulated Cells for Improving CD34$^+$ Progenitor Maintenance and Expansion

Retno Wahyu Nurhayati [1,2,*], Rafianto Dwi Cahyo [2,3], Gita Pratama [4,5], Dian Anggraini [2,6], Wildan Mubarok [2,7], Mime Kobayashi [8] and Radiana Dhewayani Antarianto [2,9]

1 Department of Chemical Engineering, Faculty of Engineering, Universitas Indonesia, Jl. Prof. Soemantri Brojonegoro, Kampus UI, Depok 16424, Indonesia
2 Stem Cells & Tissue Engineering Research Cluster, Indonesian Medical Education and Research Institute (IMERI), Faculty of Medicine, Universitas Indonesia, Jl. Salemba Raya No. 6, Central Jakarta 10430, Indonesia; rafianto_dwi_cahyo@bio.eng.osaka-u.ac.jp (R.D.C.); dian.anggraini.cu7@ms.naist.jp (D.A.); wildanmubarok@cheng.es.osaka-u.ac.jp (W.M.); radiana.dhewayani@ui.ac.id (R.D.A.)
3 Department of Biotechnology, Faculty of Engineering, 2-1 Yamadaoka, Suita, Osaka 565-0871, Japan
4 Department of Obstetric and Gynecology, Faculty of Medicine, Universitas Indonesia—Dr. Cipto Mangunkusumo General Hospital, Jl. Diponegoro No. 71, Central Jakarta 10430, Indonesia; gitapratama@yahoo.com
5 Integrated Service Unit of Stem Cell Medical Technology (IPT TK Sel Punca), Dr. Cipto Mangunkusumo General Hospital (RSCM), Jl. Diponegoro No. 71, Salemba, Central Jakarta 10430, Indonesia
6 Division of Materials Science, Graduate School of Science and Technology, Nara Institute of Science and Technology, 8916-5 Takayama, Ikoma 630-0192, Japan
7 Division of Chemical Engineering, Department of Materials Engineering Science, Graduate School of Engineering Science, Osaka University, 1-3 Machikaneyama-cho, Toyonaka, Osaka 560-8531, Japan
8 Division of Biological Science, Graduate School of Science and Technology, Nara Institute of Science and Technology, 8916-5 Takayama, Ikoma 630-0192, Japan; kobayashi.mime01@bs.naist.jp
9 Department of Histology, Faculty of Medicine, Universitas Indonesia, Jl. Salemba Raya No. 6, Central Jakarta 10430, Indonesia
* Correspondence: retno.wahyu01@ui.ac.id

Citation: Nurhayati, R.W.; Cahyo, R.D.; Pratama, G.; Anggraini, D.; Mubarok, W.; Kobayashi, M.; Antarianto, R.D. Alginate-Chitosan Microencapsulated Cells for Improving CD34$^+$ Progenitor Maintenance and Expansion. *Appl. Sci.* **2021**, *11*, 7887. https://doi.org/10.3390/app11177887

Academic Editor: Federica Boschetti

Received: 1 August 2021
Accepted: 20 August 2021
Published: 26 August 2021

Publisher's Note: MDPI stays neutral with regard to jurisdictional claims in published maps and institutional affiliations.

Abstract: Protocols for isolation, characterization, and transplantation of hematopoietic stem cells (HSCs) have been well established. However, difficulty in finding human leucocyte antigens (HLA)-matched donors and scarcity of HSCs are still the major obstacles of allogeneic transplantation. In this study, we developed a double-layered microcapsule to deliver paracrine factors from non-matched or low-matched HSCs to other cells. The umbilical cord blood-derived hematopoietic progenitor cells, identified as CD34$^+$ cells, were entrapped in alginate polymer and further protected by chitosan coating. The microcapsules showed no toxicity for surrounding CD34$^+$ cells. When CD34$^+$ cells-loaded microcapsules were co-cultured with bare CD34$^+$ cells that have been collected from unrelated donors, the microcapsules affected surrounding cells and increased the percentage of CD34$^+$ cell population. This study is the first to report the potency of alginate-chitosan microcapsules containing non-HLA-matched cells for improving proliferation and progenitor maintenance of CD34$^+$ cells.

Keywords: hematopoietic; CD34; progenitor; stem cells; microencapsulation; chitosan; alginate; proliferation; megakaryocyte

1. Introduction

Hematopoietic stem cells (HSCs) are multipotent cells capable of generating all blood components, including erythrocytes, leucocytes, and thrombocytes [1]. HSCs represent only a fraction of the cell population in umbilical cord blood, bone marrow, or peripheral blood. To isolate the HSCs from heterogenous populations, specific surface markers such as HLA-DR, CD38, CD117 (c-kit), CD45, CD133, and/or CD34 have been used [2]. One of the prominent surface markers that is widely used to isolate hematopoietic progenitor from heterogenous populations is CD34 [3,4]. $CD34^+$ cells that have enhanced progenitor activity represent a small proportion of the population [4]. Previous studies have reported that transplantation of $CD34^+$ cells successfully established multi-lineage hematopoietic engraftment [5] and improved neurobehavior of animals with stroke [6,7], brain injury [8], and cerebral palsy [9].

In clinical cases, HSCs transplantation is required for various applications, including repeated chemo- or radiotherapy for cancer patients, leukemia, lymphoma, or blood and bone marrow disorders [10]. Although the efficacy has been proven and protocol has been established, providing HLA-matched HSCs is challenging due to the diversity of HLA types and because the cells hardly propagate ex vivo.

In this study, we propose microencapsulation of non- or low-matched $CD34^+$ cells for protection against immune rejection and facilitating paracrine excretion. Microencapsulated $CD34^+$ cells can be utilized to deliver paracrine factors of unrelated $CD34^+$ cells for ex vivo expansion with a bioreactor system. For a long-term target, the microencapsulation system can be further developed to deliver paracrine factors, which provide signals to resident HSCs for regulating the homeostasis of HSCs in their niche. Two main activities that should be performed by HSCs to maintain homeostasis are cell proliferation and stemness preservation.

The novelty of this study lies in the design of a 3-dimensional (3D) scaffold for encapsulating $CD34^+$ cells and the idea for co-cultures between $CD34^+$ cells from different donors. To our knowledge, published works on 3D HSC scaffolds were generally used for ex vivo HSC culture and expansion systems and as a model for interaction between HSCs and their microenvironment [11,12]. Alginate-based capsules [13,14], 3D polyethylene terephthalate-based nanofiber [15], and collagen gel [16] have been studied as mesenchymal stem cells (MSCs) loaded scaffolds [17] and HSC-MSC micro-aggregates [16]. In these studies, biocompatibility, durability, and transplant feasibility have not always been the main considerations for designing the systems. We initially developed collagen-based scaffolds for encapsulating $CD34^+$ cells [18]. Cell viability was not affected by the collagen-based encapsulation, but it is likely that collagen promotes cell differentiation as indicated by decreasing $CD34^+$ cell population [18]. In the next step, alginate-based microcapsules were constructed [19]. In this paper, chitosan, a biocompatible polymer [20], was cross-linked with glutaraldehyde [21] and used as an outer layer to further improve alginate microcapsule stability [22]. The impact of co-culture between microencapsulated $CD34^+$ cells and bare $CD34^+$ cells from different donors was investigated. Paracrine effects by the microencapsulated $CD34^+$ cells towards the bare $CD34^+$ cells in the lower well were analyzed for cell viability and progenitor maintenance represented by the CD34 marker.

2. Materials and Methods

2.1. $CD34^+$ Cell Isolation

$CD34^+$ cell isolation from unrelated donors was conducted as described in our previous work [18]. In brief, mononuclear cells were isolated from human umbilical cord blood (UCB) by a density gradient technique with Lympoprep (STEMCELL Technologies, Vancouver, CO, Canada). UCB was collected from Dr. Cipto Mangunkusumo General Hospital after the participants were given informed consent. Ethical clearance for this study was ratified by the Ethics Committee of the Faculty of Medicine, Universitas Indonesia—Dr. Cipto Mangunkusumo General Hospital. $CD34^+$ cells were purified from mononuclear cells by EasySep™ Human Cord Blood CD34 Positive Selection Kit II (class II anti-CD34 antibody clone; STEMCELL Technologies, Vancouver, CO, Canada).

2.2. Alginate-Chitosan Microcapsule Fabrication

Microcapsules were fabricated based on our previous work [19]. Briefly, 1 part cell suspension (5×10^6 cells/mL) was mixed with 4 parts alginate solution (50 mg/mL) and then the mixture was dropped (10 µL/drop) into $CaCl_2$ solution. The alginate microcapsules were subsequently coated with glutaraldehyde cross-linked-chitosan (10 mg/mL) and neutralized with NaOH (1N). Microcapsules were washed 3 times with Mg/Ca-free phosphate-buffered saline (PBS) and tested in phenol red medium to assure neutral acidity. All reagents were purchased from Sigma-Aldrich (Gillingham, UK) unless otherwise stated. DNA staining and microscopic observation of microencapsulated cells were performed according to our published work [18].

2.3. Cytotoxicity Assay

In each experiment, 5 alginate-chitosan microcapsules were soaked in 300 μL phenol red-free RPMI medium supplemented with 10% outdated platelet lysate (Indonesian Red Cross (PMI), Jakarta, Indonesia), heparin (10 U/mL) (Fahrenheit, Indonesia), and antibiotic-antimycotic (Life Technologies, Carlsbad, CA, USA) for 72 h. The media was filtered to remove the microcapsules and subsequently used to culture $CD34^+$ cells in 96 well-plate (10^4 cells/well) (Biologix, Changzhou, China). Control cells were $CD34^+$ cells cultured in fresh media. After 24 h incubation, an equal volume of MTT solution (ThermoFisher, Waltham, MA, USA) was added directly to the culture and incubated for an additional 24 h at a refrigerated shaker (~4 °C). Insoluble formazan was dissolved with 25 μL dimethyl sulfoxide (DMSO), and the absorbance was measured at 540 nm with a spectrophotometer (Varioskan, ThermoFisher, Waltham, MA, USA).

2.4. Cell Leakage Detection Assay

Cell-loaded microcapsules were incubated in platelet lysate-supplemented RPMI media (6 microcapsules per mL media) and then incubated at 37 °C, 5% CO_2 in a fully humidified incubator. On day 8, the microcapsules were removed, and the media were collected. The detection of cell leakage from microcapsules was conducted by microscopic observation and DNA analysis of soaked media. DNA isolation was conducted according to a manufacturer's protocol (Geneaid, Taiwan) and DNA concentration was measured at 260 nm with a spectrophotometer (Biochrom, Cambridge, UK). Non-encapsulated cells (10^6 cells/mL) and fresh media were used as positive and negative controls of DNA detection, respectively.

2.5. Co-Cultures of Microencapsulated $CD34^+$ Cells with Bare $CD34^+$ Cells

$CD34^+$ cells in various concentrations (2.5, 5, or 10 $\times$ 10^4 cells/mL) were cultured in a 24 multiwell plate containing 500 μL StemSpan medium (Stem Cell Technologies) supplemented with thrombopoietin (10 ng/mL; Wako Pure Chemical, Osaka, Japan), 10% outdated platelet lysate, heparin, and antibiotic-antimycotic. In the 1st experiment set, 3 $CD34^+$ cell-loaded microcapsules were placed in each well and separated from bare cells by a polycarbonate cell culture insert (ThermoFisher, Waltham, MA, USA). In the 2nd experiment set, seeded cells (2.5 $\times$ 10^4 cells/mL) were co-cultured with 3 $\times$ 10^4 cells/well of non-encapsulated (bare) or encapsulated cells. The $CD34^+$ cells for encapsulated and non-encapsulated (bare) cells in the upper well and bare cells in the lower well from 1st and 2nd experiment sets were collected from unrelated donors. Half of the media was changed after 4 days, and the cells were harvested on day 8. Viable cell counting was performed by a dye exclusion method using trypan blue (Life Technologies, Carlsbad, CA, USA).

2.6. Flow Cytometry Analysis

Flow cytometry analysis was performed as described elsewhere [18]. In brief, the cells were stained with PE-conjugated CD34 (class III antibody clones 8G12) or FITC-conjugated CD41 antibodies (BD Biosciences, San Jose, CA, USA) for 30 min at 4 °C. The stained cells were analyzed with FACSAria III flow cytometer (BD Biosciences, San Jose, CA, USA).

2.7. Statistical Analysis

Data were presented as mean values $\pm$ standard errors from triplicate experiments. Data sets were analyzed using one-way ANOVA, with post-hoc analysis using a paired t-test. The value of $p < 0.05$ was considered as a significant difference.

3. Results and Discussion

3.1. Fabrication of a Durable Double-Layered Microcapsule

A double-layered alginate-chitosan microcapsule was fabricated to encapsulate CD34$^+$ progenitor cells (Figure 1A,B). The coating is intended to protect the cells from immune cell intrusion and, at the same time, to be permeable enough to allow nutrient and gas exchange or paracrine secretion (Figure 1A). The core is alginate-entrapped cells, and the outer layer consists of chitosan cross-linked with glutaraldehyde (Figure 1B,D). The glutaraldehyde cross-linked chitosan coating improved alginate microcapsule stability (Figure 1C). Based on microscopic observations, the alginate-chitosan microcapsules retained its spherical shape for at least 12 days in culture media. The DNA analysis was conducted to confirm no cell leakage from microcapsules into media. Non-encapsulated cells and fresh media were used as positive and negative controls of DNA detection, respectively. DNA concentration was zero for microcapsule-soaked media (n = 3) and fresh media, whereas DNA concentration of positive control (10^6 cells) was 26.3 ng/µL. Chelating agents (i.e., phosphate, citrate, lactate) and non-gelling cations (i.e., Na^+, Mg^{2+}) contained in the media are known to destabilize calcium alginate gel [23]. Non-coated calcium alginate microcapsules typically break down after 3-day immersion in common saline [19]. Our experimental data suggested the alginate-chitosan microcapsules were able to provide a physical barrier in the employed culture condition.

Next, the cytotoxicity effect of alginate-chitosan microcapsules was analyzed by an indirect MTT (3-[4,5-dimethylthiazol-2-yl]-2,5 diphenyl tetrazolium bromide) assay [24] after microcapsules were immersed in media for 3 days. There was no significant difference of cell viabilities cultured in media with or without microcapsules (Figure 1E). This result suggests that the microcapsules do not pose a cytotoxic effect on surrounding cells.

3.2. Paracrine Effects of Microencapsulated CD34$^+$ Cells

To reconstitute the communication condition in HSC niche, co-culture experiments of microencapsulated CD34$^+$ cells and bare CD34$^+$ cells were set up. The experiments were performed in a static bath culture to facilitate paracrine factors excreted from microencapsulated cells. The microcapsules were placed on the top part, while the non-encapsulated/bare CD34$^+$ cells were in the lower part. A polycarbonate insert was used for easy cell harvest (Figure 2A). Initially, we prepared 3 microcapsules/well that contains a total of 6×10^3 cells/well (Donor 1; upper well) for co-culture with 2.5, 5, or 10×10^4 cells/mL (Donor 2, lower well). There was a significant increase in CD34$^+$ cell population of 2.5×10^4 cells (lower well) after 8-day co-culturing, in comparison with those without microencapsulated cells (data not shown). However, the CD34$^+$ ratio did not improve in higher seeding density cultures. With the same treatment, no impact was seen in cell proliferation of all seeding densities (Figure S1). Therefore, we increased the cell loading capacity of each of 3 microcapsules to contain 10^4 cells/microcapsule (total 3×10^4 cells/well) (Donor 3; upper well) for co-culture with 2.5, 5, and 10×10^4 cells/mL (Donor 4; lower well).

As seen in Figure 2B, microcapsules containing 3×10^4 cells per well improved the viable cell densities of bare CD34$^+$ cells (lower well) seeded in 2.5 and 5×10^4 cells/mL, but made no difference at a higher seeding density. At seeding density of 2.5×10^4 cells/mL and 5×10^4 cells/mL, cell concentrations of those with microencapsulated cells increased from $2.7 \pm 0.3 \times 10^4$ cells/mL to $4.7 \pm 0.6 \times 10^4$ cells/mL and from $3.5 \pm 0.5 \times 10^4$ cells/mL to $7.1 \pm 1.6 \times 10^4$ cells/mL, respectively. We expected that for seeding density 10×10^4 cells/mL required a higher dosage of microencapsulated cells.

Typically, freshly isolated CD34$^+$ cells from UCB have a purity of $93.0 \pm 0.2\%$ (indicated as a striped line in Figure 2C). After 8-day culture, the CD34 positive cell percentage dropped to 83.2 ± 1.7, 82.7 ± 2.0 and $88.4 \pm 0.5\%$ in CD34$^+$ cells seeded at 2.5, 5, 10×10^4 cells/mL, respectively. Interestingly, CD34$^+$ ratio increased to 89.3 ± 1.2, 96.0 ± 0.1 and $94.4 \pm 0.4\%$ in cultures treated with microencapsulated cells. These results suggested that microencapsulated cells potentially maintained CD34 expression in in vitro culture.

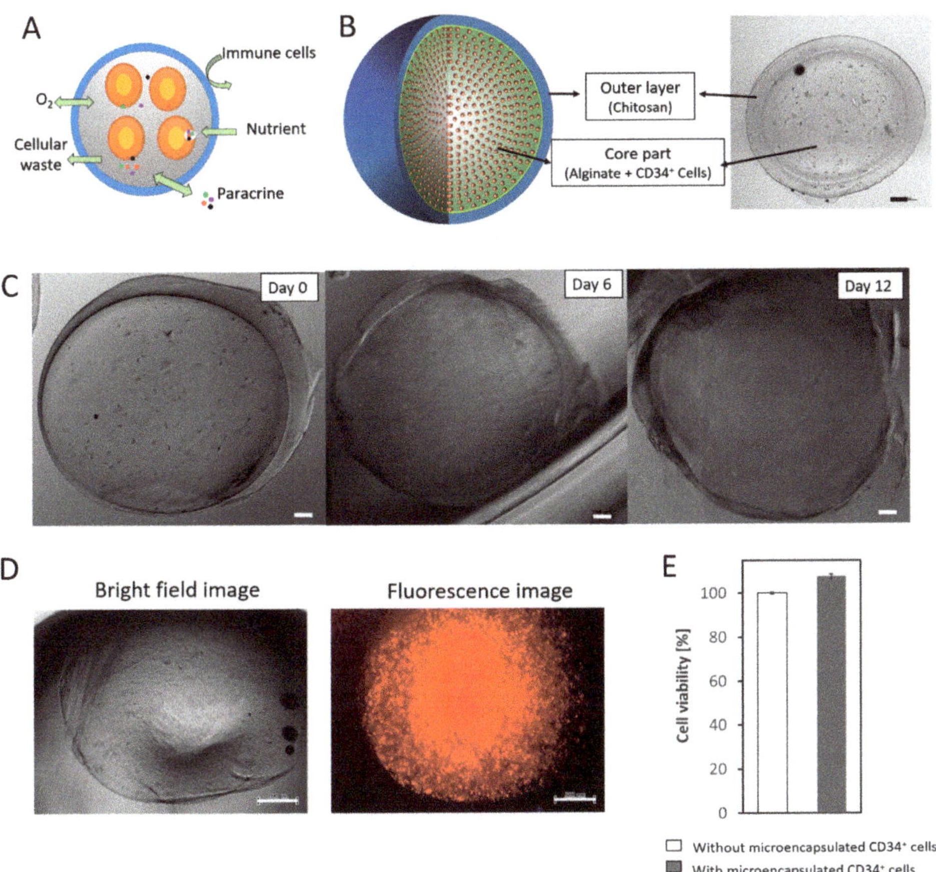

Figure 1. (**A**) Illustration of microencapsulation with a semi-permeable barrier. (**B**) Schematic and microscopic image of double-layered alginate-chitosan microcapsule. (**C**) Stability of alginate-chitosan microcapsules in culture media. (**D**) Microscopic images of cell-loaded microcapsule after DNA staining. Scale bars in (**B–D**) represent 100 and 200 μm, respectively. (**E**) Cytotoxicity of alginate-chitosan microcapsules for external CD34$^+$ cells.

3.3. Importance of Microenvironment for CD34$^+$ Progenitor Maintenance

Hematopoiesis, including self-renewal and differentiation, is highly regulated by the bone marrow microenvironment. Autocrine/paracrine factors such as growth factors, cytokines, and chemokines are secreted by early and differentiated hematopoietic cells, may play important roles as chemo-attractants for other hematopoietic cells (accessory cells, facilitating cells, etc.), stimulate the secretion of other regulatory molecules, modulate the expression of adhesion molecules on the cell surface (e.g., adhesion and homing), and/or influence the survival of the hematopoietic cells [25].

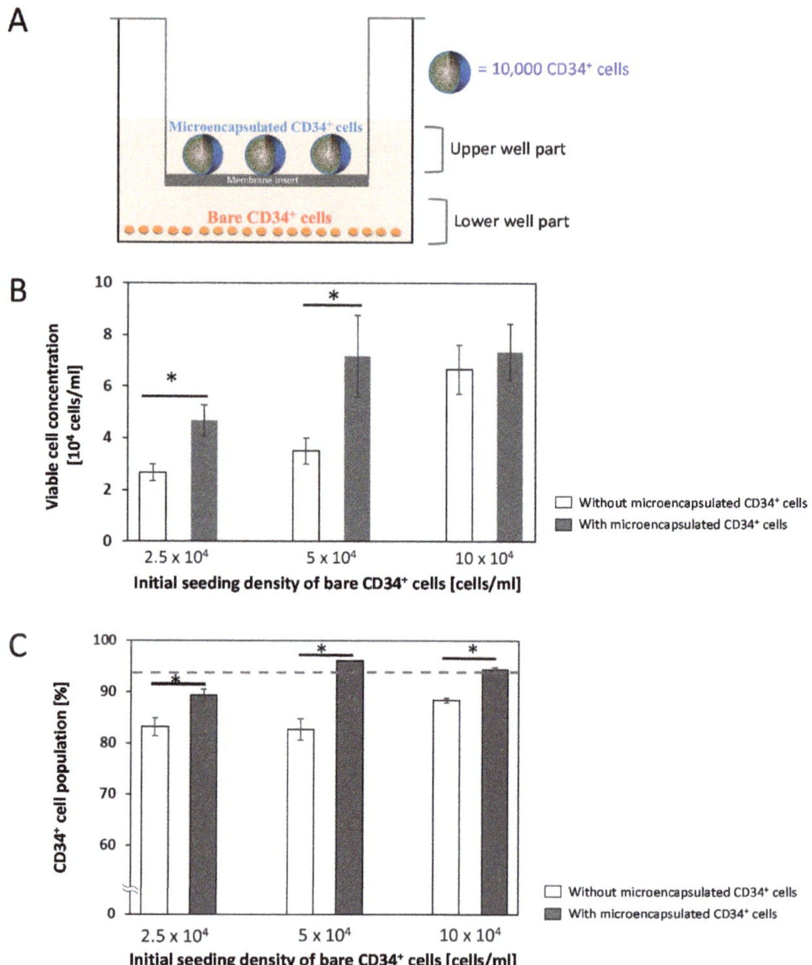

Figure 2. (**A**) Illustration of co-culture system of CD34$^+$ cell-loaded microcapsules (3×10^4 cells/well) with bare CD34$^+$ cells from unrelated donors. (**B**) Viable cell concentrations of bare cells (lower well) on day 8. (**C**) Ratio of CD34$^+$ cell population (lower well) on day 8. Striped line in C indicates percentage of CD34$^+$ cell population on day 0. Asterisks (*) in (**B**,**C**) indicate $p < 0.05$.

In a recent study, a micro-cavity platform was developed for isolating HSC to grow solely or in certain cell numbers thus that it can be used to investigate the impact of autocrine and paracrine signals in HSC culture [26]. The hypothesis was autocrine signals are predominantly involved in maintaining the quiescent state of HSC in single-cell niches [26]. However, it was implied that single-cell or multi-cell cultures were unable to preserve stemness state although various growth factors were introduced (stem cell factor, thrombopoietin, and FLT3L). The experimental results showed that decreasing in CD34$^+$ ratio is more likely to happen in lower cell density. The CD34$^+$ cell population was preserved by co-culturing with microencapsulated CD34$^+$ cells. Moreover, cell expansion increased significantly in the presence of microencapsulated CD34$^+$ cells (Figure S2). The number of loaded cells had a significant impact on the effectiveness of microencapsulated CD34$^+$ cells.

The paracrine release from microencapsulated cells was hypothesized to play an important role in promoting the proliferation and maintenance of cultured CD34$^+$ cells.

Additional experiments were conducted to address this hypothesis. The cells (Donor 6; lower well) were co-cultured with equivalent numbers of bare or encapsulated cells from an unrelated donor (Donor 5; upper well). On day 8, the cell viabilities of bare cells (lower well) were 94.4 ± 3.2; 91.7 ± 4.8 and 100 ± 0.0% for monoculture, co-culture with bare cells, and co-culture with microencapsulated cells. It was noticed that co-cultures with encapsulated or bare cells from the unrelated donor could improve the cell expansion (Figure S3A). Moreover, co-cultures produced higher percentages of $CD34^+$ cells than of monoculture (Figure S3B). These results suggested that co-culture with encapsulated cells, isolated from an unrelated donor, have beneficial effects for $CD34^+$ cell expansion and progenitor maintenance.

In the present study, thrombopoietin, which can promote cell proliferation and megakaryocytic differentiation [27], was added because platelet lysate alone could not provide enough growth factors for $CD34^+$ cell survival and proliferation in vitro (Figure S4). CD41 is expressed throughout megakaryocytic lineage and thus considered as a representative marker for megakaryotic differentiation [3,28]. It was found that bare and encapsulated cells have opposite effects on CD41 expression of co-cultured cells (Figure S5). Co-culture with bare cells from different donors suppressed CD41 expression of cultured cells (lower well). On the contrary, when the equivalent cells were encapsulated with alginate-chitosan, it could promote CD41 expression of cultured cells (lower well). To our knowledge, this is the first time such a paracrine effect is reported.

There are two possible mechanisms: (1) direct effect in which chitosan and/or alginate interact with thrombopoietin and subsequently promote megakaryocytic differentiation of external cells or (2) indirect effect in which chitosan and/or alginate interact with thrombopoietin or encapsulated cells and subsequently cause the entrapped cells to release paracrine factors that promote megakaryocytic differentiation of external cells. Further studies are needed to understand the detailed mechanism.

Encapsulation for transplantation was initially developed to improve allogeneic pancreas islet transplantation for diabetic patients [29]. Semi-permeable membrane provides mechanical protection and immune rejection for transplanted islets while nutrients and cellular metabolites freely diffuse through the membrane [30]. Several clinical trials have been performed, and the follow-up evaluation showed a safe and promising result of alginate-based microencapsulated islets for diabetic patients [30]. Inspired by these findings, we developed $CD34^+$ cells-loaded microcapsule to demonstrate paracrine factor delivery. The microcapsule possessed no toxicity for external cells. The 3D capsules could restrict cell growth of mesenchymal stem cells (MSCs) [31], myoblast and fibroblast cells [32], which was postulated due to limited nutrients/oxygen exchange. Limited cell growth is beneficial to prevent overpopulation inside the capsule as long as the cells are still capable of releasing paracrine factors. In addition, microencapsulation can be an advantage in facilitating scaled-up co-culture systems for HSC or progenitor expansion with bioreactors (Figure S6).

HSC homeostasis maintenance required paracrine factors secreted by niche cells to supply or feed the right signal for HSCs [33]. A study reported encapsulation of several cell types, including human amniotic-MSCs, umbilical cord-MSCs, and fibroblasts, to maintain $CD34^+$ cell expansion in vitro [31]. Their finding indicated that encapsulated MSCs or fibroblasts do not benefit $CD34^+$ cells proliferation [31]. Interestingly, improved cell proliferation of $CD34^+$ cells by co-culturing with osteosarcoma cell line has been reported [34]. Instead of employing other cells, our study used $CD34^+$ cells collected from unrelated donors as the source of paracrine factors required for hematopoietic progenitor maintenance. This system can be utilized to deliver paracrine factors for ex vivo HSC or progenitor expansion with bioreactor and further developed to support initial engraftment of allogeneic HSC transplantation.

4. Conclusions

A non-toxic double-layered microcapsule has been successfully fabricated with alginate-chitosan to encapsulate human $CD34^+$ progenitor cells. The microcapsules did not pose any toxicity to external cells. Cell proliferation and progenitor maintenance were significantly improved when the cells from unrelated donors were cultured with CD34$^+$ cell-loaded microcapsules. The thrombopoietin-induced megakaryocytic differentiation was amplified when the cells were co-cultured with cell-loaded microcapsules, while it was repressed when co-cultured with non-encapsulated cells. Our study is the first proof-of-concept that microencapsulation can be utilized to deliver paracrine factors for ex vivo expansion of HLA-unmatched CD34$^+$ progenitors.

Supplementary Materials: The following are available online at https://www.mdpi.com/article/10.3390/app11177887/s1, Figure S1: A. Illustration of co-culture system of microencapsulated CD34$^+$ cells (6×10^3 cells/well) with bare CD34$^+$ cells from unrelated donors. B. Viable cell concentration of bare cells after co-culturing for 8 days, Figure S2: Fold expansions of CD34$^+$ cells (lower well) on day 8 post-culturing with and without cell-loaded microcapsules (3×10^4 cells/well), Figure S3: A. Viable cell concentration of bare cells (lower well) on day 8. B. Ratio of CD34$^+$ cell population on day 8. Bare cells and microencapsulated cells (Donor 5, upper well) were 3.0×10^4 cells/well. Initial seeding of bare cells (Donor 6; lower well) was 2.5×10^4 cells/ml. Donor 5 and 6 were unrelated patients. Figure S4: Effect of thrombopoietin (TPO) for CD34$^+$ cells cultured on platelet-lysate-(PL)-supplemented media, Figure S5: The CD41$^+$ cell population of bare cells (lower well) after 8-day monoculture or co-cultures in thrombopoietin-supplemented media; Illustration of co-culture systems with microcapsule-bioreactor and membrane insert.

Author Contributions: R.W.N.: Conceptualization, Methodology, Investigation, Data Curation, Writing, Visualization, Supervision. R.D.C.: Methodology, Investigation, Resources. G.P.: Resources. D.A.: Formal analysis, Validation. W.M.: Formal analysis, Writing–Review and Editing. M.K.: Writing–Review and Editing. R.D.A.: Resources. All authors have read and agreed to the published version of the manuscript.

Funding: This work was partially supported by Q1Q2 Scheme Research Grant (Hibah Publikasi Artikel di Jurnal Internasional Kuartil Q1 dan Q2 (Q1Q2) No. NKB-0232/UN2.R3.1/HKP.05.00/2019) and P5 Grant (Program Pendanaan Perancangan dan Pengembangan Purwarupa) No. PKS-13/UN2.INV/HKP.05/2020 from Universitas Indonesia awarded to Dr. Retno Wahyu Nurhayati.

Institutional Review Board Statement: Ethical clearance for this study was ratified by the Ethics Committee of the Faculty of Medicine, Universitas Indonesia—Dr. Cipto Mangunkusumo General Hospital.

Informed Consent Statement: Informed consent was obtained from all subjects involved in the study.

Data Availability Statement: All data generated or analysed during this study are included in this published article and its supplementary information files.

Acknowledgments: The authors thank Cipto Mangunkusumo General Hospital (RSCM) for facilitating UCB collection, Sinha Haelamani and Elizabeth Agustina for technical support.

Conflicts of Interest: The authors declare no conflict of interest.

References

1. Akashi, K.; Traver, D.; Miyamoto, T.; Weissman, I.L. A clonogenic common myeloid progenitor that gives rise to all myeloid lineages. *Nature* **2000**, *404*, 193–197. [CrossRef]
2. Sidney, L.E.; Branch, M.J.; Dumphy, S.E.; Dua, H.S. Concise Review: Evidence for CD34 as a Common Marker for Diverse Progenitors. *Stem Cells* **2014**, *32*, 1380–1389. [CrossRef]
3. Nurhayati, R.W.; Ojima, Y.; Taya, M. Recent developments in ex vivo platelet production. *Cytotechnology* **2016**, *68*, 2211–2221. [CrossRef] [PubMed]
4. Engelhardt, M.; Lübbert, M.; Guo, Y. CD34$^+$ or CD34$^-$: Which is the more primitive? *Leukemia* **2002**, *16*, 1603–1608. [CrossRef] [PubMed]
5. Gao, Z.; Fackler, M.J.; Leung, W.; Lumkul, R.; Ramirez, M.; Theobald, N.; Malech, H.L.; Civin, C.I. Human CD34+ cell preparations contain over 100-fold greater NOD/SCID mouse engrafting capacity than do CD34- cell preparations. *Exp. Hematol.* **2001**, *29*, 910–921. [CrossRef]
6. Chen, J.; Sanberg, P.R.; Li, Y.; Wang, L.; Lu, M.; Willing, A.E.; Sanchez-Ramos, J.; Chopp, M. Intravenous administration of human umbilical cord blood reduces behavioral deficits after stroke in rats. *Stroke* **2001**, *32*, 2682–2688. [CrossRef]

7. Taguchi, A.; Soma, T.; Tanaka, H.; Kanda, T.; Nishimura, H.; Yoshikawa, H.; Tsukamoto, Y.; Iso, H.; Fujimori, Y.; Stern, D.M.; et al. Administration of CD34$^+$ cells after stroke enhances neurogenesis via angiogenesis in a mouse model. *J. Clin. Investig.* **2004**, *114*, 330–338. [CrossRef]
8. Verina, T.; Fatemi, A.; Johnston, M.V.; Comi, A.M. Pluripotent possibilities: Human umbilical cord blood cell treatment after neonatal brain injury. *Pediatr. Neurol.* **2013**, *48*, 346–354. [CrossRef]
9. Chang, Y.; Lin, S.; Li, Y.; Liu, S.; Ma, T.; Wei, W. Umbilical cord blood CD34$^+$ cells administration improved neurobehavioral status and alleviated brain injury in a mouse model of cerebral palsy. *Child's Nerv. Syst.* **2021**, *37*, 1–9. [CrossRef]
10. Gratwohl, A.; Pasquini, M.C.; Aljurf, M.; Atsuta, Y.; Baldomero, H.; Foeken, L.; Gratwohl, M.; Bouzas, L.F.; Confer, D.; Frauendorfer, K.; et al. One million haemopoietic stem-cell transplants: A retrospective observational study. *Lancet Haematol.* **2015**, *2*, e91–e100. [CrossRef]
11. Choi, J.S.; Mahadik, B.; Harley, B.A.C. Engineering the hematopoietic stem cell niche: Frontiers in biomaterial science. *Biotechnol. J.* **2015**, *10*, 1529–1545. [CrossRef] [PubMed]
12. Nies, C.; Gottwald, E. Artificial Hematopoietic Stem Cell Niches-Dimensionality Matters. *Adv. Tissue Eng. Regen. Med.* **2017**, *2*, 00042.
13. Yuan, Y.; Tse, K.-T.; Sin, F.W.-Y.; Xue, B.; Fan, H.-H.; Xie, Y. Ex vivo amplification of human hematopoietic stem and progenitor cells in an alginate three-dimensional culture system. *Int. J. Lab. Hematol.* **2011**, *33*, 516–525. [CrossRef] [PubMed]
14. Tarunina, M.; Hernandez, D.; Kronsteiner-Dobramysl, B.; Pratt, P.; Watson, T.; Hua, P.; Gullo, F.; Van Der Garde, M.; Zhang, Y.; Hook, L.; et al. A Novel High-Throughput Screening Platform Reveals an Optimized Cytokine Formulation for Human Hematopoietic Progenitor Cell Expansion. *Stem Cells Dev.* **2016**, *25*, 1709–1720. [CrossRef] [PubMed]
15. Li, Y.; Ma, T.; Kniss, D.A.; Yang, S.-T.; Lasky, L.C. Human Cord Cell Hematopoiesis in Three-Dimensional Nonwoven Fibrous Matrices: In Vitro Simulation of the Marrow Microenvironment. *J. Hematother. Stem Cell Res.* **2001**, *10*, 355–368. [CrossRef] [PubMed]
16. Leisten, I.; Kramann, R.; Ferreira, M.S.V.; Bovi, M.; Neuss, S.; Ziegler, P.; Wagner, W.; Knüchel, R.; Schneider, R.K. 3D co-culture of hematopoietic stem and progenitor cells and mesenchymal stem cells in collagen scaffolds as a model of the hematopoietic niche. *Biomaterials* **2012**, *33*, 1736–1747. [CrossRef]
17. Cook, M.M.; Futrega, K.; Osiecki, M.; Kabiri, M.; Kul, B.; Rice, A.; Atkinson, K.; Brooke, G.; Doran, M. Micromarrows—Three-Dimensional Coculture of Hematopoietic Stem Cells and Mesenchymal Stromal Cells. *Tissue Eng. Part C* **2012**, *18*, 319–328. [CrossRef] [PubMed]
18. Nurhayati, R.W.; Antarianto, R.D.; Pratama, G.; Rahayu, D.; Mubarok, W.; Kobayashi, M.; Hutabarat, M. Encapsulation of human hematopoietic stem cells with a biocompatible polymer. *AIP Conf. Proc.* **2019**, *2092*, 020011.
19. Nurhayati, R.W.; Cahyo, R.D.; Alawiyah, K.; Pratama, G.; Agustina, E.; Antarianto, R.D.; Prijanti, A.R.; Mubarok, W.; Rahyussalim, A.J. Development of Double-Layered Alginate-Chitosan Hydrogels for Human Stem Cell Microencapsulation. *AIP Conf. Proc.* **2019**, *2193*, 020004.
20. Kong, X.; Xu, W. Biodegradation and biocompatibility of a degradable chitosan vascular prosthesis. *Int. J. Clin. Exp. Med.* **2015**, *8*, 3498–3505.
21. Fu, J.; Yang, F.; Guo, Z. The chitosan hydrogels: From structure to function. *New J. Chem.* **2018**, *42*, 17162–17180. [CrossRef]
22. Batubara, I.; Rahayu, D.; Mohamad, K.; Prasetyaningtyas, W.E. Leydig Cells Encapsulation with Alginate-Chitosan: Optimization of Microcapsule Formation. *J. Encapsulation Adsorpt. Sci.* **2012**, *2*, 15–20. [CrossRef]
23. Thu, B.; Smidsrød, O.; Skjåk-Bræk, G. Alginate gels—Some structure-function correlations relevant to their use as immobilization matrix for cells. *Immobil. Cells* **1996**, *11*, 19–30.
24. van Meerloo, J.; Kaspers, G.J.L.; Cloos, J. Cell sensitivity assays: The MTT assay. *Methods Mol. Biol.* **2011**, *73*, 237–245.
25. Janowska-Wieczorek, A.; Majka, M.; Ratajczak, J.; Ratajczak, M.Z. Autocrine/Paracrine Mechanisms in Human Hematopoiesis. *Stem Cells.* **2001**, *19*, 99–107. [CrossRef] [PubMed]
26. Müller, E.; Wang, W.; Qiao, W.; Bornhäuser, M.; Zandstra, P.W.; Werner, C.; Pompe, T. Distinguishing autocrine and paracrine signals in hematopoietic stem cell culture using a biofunctional microcavity platform. *Sci. Rep.* **2016**, *6*, 31951. [CrossRef] [PubMed]
27. Ojima, Y.; Duncan, M.T.; Nurhayati, R.W.; Taya, M.; Miller, W. Synergistic effect of hydrogen peroxide on polyploidization during the megakaryocytic differentiation of K562 leukemia cells by PMA. *Exp. Cell Res.* **2013**, *319*, 2205–2215. [CrossRef] [PubMed]
28. Deutsch, V.R.; Tomer, A. Megakaryocyte development and platelet production. *Br. J. Haematol.* **2006**, *134*, 453–466. [CrossRef]
29. Sakata, N.; Sumi, S.; Yoshimatsu, G.; Goto, M.; Egawa, S.; Unno, M. Encapsulated islets transplantation: Past, present and future. *World J. Gastrointest. Pathophysiol.* **2012**, *3*, 19–26. [CrossRef] [PubMed]
30. Tuch, B.E.; Keogh, G.W.; Williams, L.J.; Wu, W.; Foster, J.L.; Vaithilingam, V.; Philips, R. Safety and Viability of Microencapsulated Human Islets Transplanted Into Diabetic Humans. *Diabetes Care* **2009**, *32*, 1887–1889. [CrossRef]
31. Pan, X.; Sun, Q.; Cai, H.; Gao, Y.; Tan, W.; Zhang, W. Encapsulated feeder cells within alginate beads for ex vivo expansion of cord blood-derived CD34+ cells. *Biomater. Sci.* **2016**, *4*, 1441–1453. [CrossRef] [PubMed]
32. del Burgo, L.S.; Ciriza, J.; Noguera, A.E.; Illa, X.; Cabruja, E.; Orive, G.; Hernandez, R.M.; Villa, R.; Pedraz, J.L.; Alvarez, M. 3D Printed porous polyamide macrocapsule combined with alginate microcapsules for safer cell-based therapies. *Sci. Rep.* **2018**, *8*, 1–14. [CrossRef]

33. Lee-Thedieck, C.; Spatz, J.P. Artificial Niches: Biomimetic Materials for Hematopoietic Stem Cell Culture. *Macromol. Rapid Commun.* **2012**, *33*, 1432–1438. [CrossRef] [PubMed]
34. Rochet, N.; Leroy, P.; Far, D.F.; Ollier, L.; Loubat, A.; Rossi, B. CAL72: A human osteosarcoma cell line with unique effects on hematopoietic cells. *Eur. J. Haematol.* **2003**, *70*, 43–52. [CrossRef] [PubMed]

 applied sciences

MDPI

Article

Assessment of Two Commonly used Dermal Regeneration Templates in a Swine Model without Skin Grafting †

Wiebke Eisler [1,*], Jan-Ole Baur [2], Manuel Held [1], Afshin Rahmanian-Schwarz [3], Adrien Daigeler [1] and Markus Denzinger [4]

1 Department of Hand, Plastic, Reconstructive and Burn Surgery, BG Klinik Tübingen, Eberhard Karls University Tübingen, 72076 Tübingen, Germany; manuelheld@hotmail.com (M.H.); adaigeler@bgu-tuebingen.de (A.D.)
2 Department of Dermatology, University Hospital Erlangen, Friedrich-Alexander-University Erlangen-Nuernberg, 91054 Erlangen, Germany; jan-ole.baur@gmx.de
3 Department of Plastic, Hand, Reconstructive and Aesthetic Surgery, Hand Surgery, Traunstein Hospital, Ludwig-Maximilians-Universität München, 80539 Munich, Germany; a.rahmanian@web.de
4 Department of Pediatric Surgery, Klinik St. Hedwig, University Medical Center Regensburg, 93049 Regensburg, Germany; m.denzinger@t-online.de
* Correspondence: w.v.petersen@gmx.de; Tel.: +49-7071-6060
† The work was performed at the University of Tübingen, Germany.

Abstract: In the medical care of partial and full-thickness wounds, autologous skin grafting is still the gold standard of dermal replacement. In contrast to spontaneous reepithelializing of superficial wounds, deep dermal wounds often lead to disturbing scarring, with cosmetically or functionally unsatisfactory results. However, modern wound dressings offer promising approaches to surface reconstruction. Against the background of our future aim to develop an innovative skin substitute, we investigated the behavior of two established dermal substitutes, a crosslinked and a non-crosslinked collagen biomatrix. The products were applied topically on a total of 18 full-thickness skin defects paravertebrally on the back of female Göttingen Minipigs—six control wounds remained untreated. The evaluation was carried out planimetrically (wound closure time) and histologically (neoepidermal cell number and epidermis thickness). Both treatment groups demonstrated significantly faster reepithelialization than the controls. The histologic examination verified the highest epidermal thickness in the crosslinked biomatrix-treated wounds, whereas the non-crosslinked biomatrix-treated wounds showed a higher cell density. Our data presented a positive influence on epidermal regeneration with the chosen dermis substitutes even without additional skin transplantation and, thus, without additional donor site morbidity. Therefore, it can be stated that the single biomatrix application might be used in a clinical routine with small wounds, which needs to be investigated further in a clinical setting to determine the size and depths of a suitable wound bed. Nevertheless, currently available products cannot solely achieve wound healing that is equal to or superior to autologous tissue. Thus, the overarching aim still is the development of an innovative skin substitute to manage surface reconstruction without additional skin grafting.

Keywords: animal model; dermal regeneration; Integra; Matriderm; skin substitute; wound healing

Citation: Eisler, W.; Baur, J.-O.; Held, M.; Rahmanian-Schwarz, A.; Daigeler, A.; Denzinger, M. Assessment of Two Commonly used Dermal Regeneration Templates in a Swine Model without Skin Grafting. *Appl. Sci.* **2022**, *12*, 3205. https://doi.org/10.3390/app12063205

Academic Editors: Francesca Silvagno and Jongsung Lee

Received: 9 February 2022
Accepted: 17 March 2022
Published: 21 March 2022

1. Introduction

In the human body, the skin is the largest organ and serves as a barrier to the external world. The epidermis is a stratified epithelium with proliferating basal and differentiated suprabasal keratinocytes. The dermis consists of an extracellular matrix with interwoven collagen fibrils, some elastin and glycosaminoglycan, to give physical strength and flexibility to the skin, and fibroblasts. These are able to produce remodeling enzymes, which are important for wound healing (e.g., collagenases and proteases). Skin injuries through trauma, disease, burns, surgery, or even chronic wounds can have drastic effects. In contrast to spontaneous reepithelializing of superficial wounds, deep dermal wounds often lead to delayed healing, disturbing scarring and cosmetically or functionally unsatisfactory results.

The ultimate goal in plastic, reconstructive and burn surgery is to produce skin analogues similar to natural skin. To achieve better outcomes with a lower risk of mortality and better functional results, early permanent wound closure is recommended. Autologous skin grafting is still the gold standard of dermal replacement in partial and full-thickness wounds. The demand for modern skin substitutes is high [1–3]. Various artificial skin replacement products have been developed recently, and their use is increasingly widespread in clinical practice [3–5]. The scaffolds must meet certain requirements, besides clinical effectiveness, easy handling and safe application to the patient. Other necessary scaffold properties are:

- Biocompatibility and biodegradability;
- Ability to guide regenerative skin elements;
- Similarity to the physical strength and flexibility of normal skin;
- The 3D-matrix of naturally existing substances of the human body.

At present, acellular skin substitutes produced through lyophilization and phase separation techniques are the most convincing imitation of the extracellular matrix of the skin [6]. Many researchers are pursuing new approaches to tissue engineering. Some currently available products offer promising properties such as protecting the wound from fluid loss and microbial invasion. They support dermal cell migration and neoangiogenesis and reduce the development of scar tissue [7]. Against the background of our future aim to develop an innovative skin substitute, we investigated the behavior of two established dermal substitutes, a crosslinked and a non-crosslinked collagen biomatrix in this study. The products were investigated in full-thickness skin defects as a promoter for epithelialization without additional skin grafts as an alternative, novel approach to the originally intended use in combination with skin grafting and directly compared with a particular focus on neodermal formation.

2. Materials and Methods

In this trial, full-thickness skin defects were generated paravertebrally in female Göttingen minipigs due to the physiologic concordance to human skin [8,9]. The animals were provided from Ellegaard Göttingen Minipigs A/S, Dalmose, Denmark, aged 39 weeks (±12 days) and weighed 22.6 kg (±1.4 kg). Animals were treated according to the German Law on the Protection of Animals, and the study was performed with permission from the local Animal Welfare Committee (approval code AT 1/12).

On the first day of the study, the Göttingen minipigs were anesthetized and their backs were shaved, surgically disinfected in a standardized manner and the outline of the wounds was tattooed. Eighteen full-thickness skin defects (2.0 cm diameter, 0.6 cm depth) were created, and a single dressing was applied with either a crosslinked biomatrix ($n = 6$) or a non-crosslinked biomatrix ($n = 6$). Control wounds remained untreated ($n = 6$). The crosslinked biomatrix (Bilayered Integra® by Integra Life Sciences Corporation, Plainsboro, NJ, USA) is an acellular crosslinked collagen type I and chondroitin-6-sulphate based bilayered biomatrix with a polysiloxane polymer layer. It is produced from bovine tendon and shark glycosaminoglycan. The silicone was left on the wound for seven days. The non-crosslinked biomatrix (Matriderm® 1.00 mm by Dr. Suwelack Skin & Health Care AG, Billerbeck, Germany) is based on a compound from lyophilized collagen type I, III, V and α-elastin in a decellularized non-crosslinked biomatrix of bovine origin. The wound dressings were applied to the wounds in a randomized manner with a standardized separation distance of 6.0 cm sealed with adhesive foil (Smith & Nephew Orthopaedics GmbH, Tuttlingen, Germany) to protect the implants, avoid cross-contamination and prevent bacterial contamination. Additionally, the minipigs wore customized minipig jackets (Ellegaard Minipig Jacket Large Full Body, Lomir Biomedical Inc., Notre Dame de L'Ille Perrot, QC, Canada), as previously described [10–13].

Bandage renewal and photographic documentation were performed every second day using a digital camera with a tripod to determine the exact distance. The evaluation was carried out planimetrically by importing the photos into Adobe Photoshop (Adobe Systems Inc., San Jose, CA, USA) and calculating the percentage of epithelialization during the study period of 21 days [14]. After 20 days, original circular wound areas were excised, and samples were processed for histopathological evaluation. Histological slices were stained with hematoxylin and eosin to analyze the resulting neodermal thickness and cell density. The thickness of the neoepidermis was determined repetitively from the basal layer to the stratum corneum of the epidermis, and the number of keratinocytes was quantified within a rectangular area of 100 × 50 µm (5 mm^2), both in three different sections with an interval of 100 µm. Pictures were taken with a digital microscope camera (AxioCam ERc 5s, Carl Zeiss Microscopy GmbH, Jena, Germany) connected to a Zeiss microscope (Axio Observer.Z1, Carl Zeiss Microscopy GmbH, Jena, Germany) with the ZEN blue edition (2011) microscopy software (Carl Zeiss Microscopy GmbH, Jena, Germany). The study was carried out after approval of the local Animal Welfare Committee in accordance with the German Animal Protection Act. Statistical evaluation was performed using SPSS software version 20.0. The results were analyzed with the Wilcoxon rank-sum test with a defined p-value ≤ 0.05.

3. Results

All templates were examined macroscopically and histologically, and upon completion of the study, epidermal coverage was complete without any rejection reaction in any animal (Figure 1).

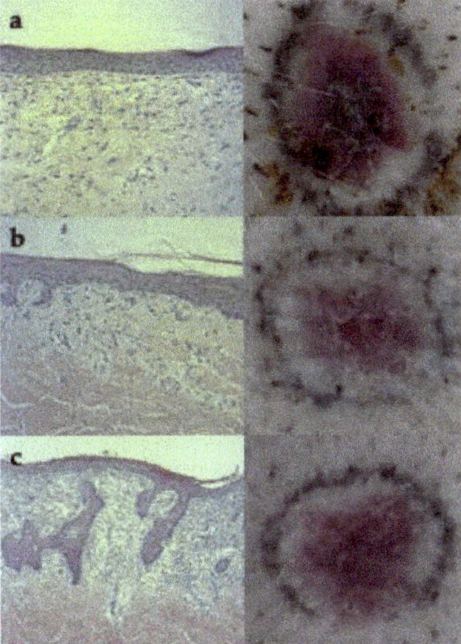

Figure 1. Macroscopic and microscopic images from tissue stained with hematoxylin and eosin of the excised former sore center in a representative (**a**) untreated wound and a wound treated with (**b**) a crosslinked and (**c**) a non-crosslinked biomatrix at 20 days after a single dressing application. All biomaterials presented a continuous epidermal layer and complete wound closure.

3.1. Planimetric Evaluation

Wound closure time decreased significantly in the treatment groups. For the control wounds, a complete epithelialization was observed after 13.50 ± 1.19 days. Planimetric analysis of wounds treated with the crosslinked biomatrix showed complete wound closure after an average of 10.00 ± 1.15 days. The non-crosslinked biomatrix-treated wounds were closed after 10.67 ± 0.94 days. Thus, the two treatment groups showed significantly faster wound closure than the controls. The mean time for complete wound closure was accelerated by 3.50 days ($p < 0.0001$) and 2.83 days ($p = 0.0002$), respectively (Figure 2).

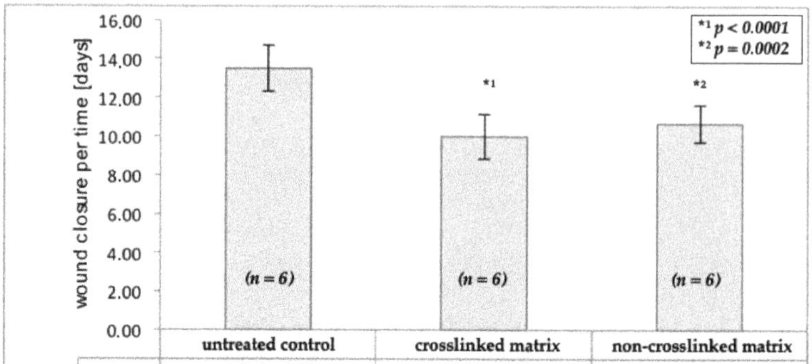

*1 $p < 0.0001$
*2 $p = 0.0002$

wound closure per time [days]

	untreated control (n = 6)	crosslinked matrix (n = 6) *1	non-crosslinked matrix (n = 6) *2
average	13.50 ± 1.19	10.00 ± 1.15	10.67 ± 0.94

Figure 2. Time in days to complete epithelialization of control wounds and treated wounds with the crosslinked and non-crosslinked biomatrices. The wound closure was significantly increased in the treatment groups. * Statistically significant results.

3.2. Histological Analysis

Further, the evaluation was carried out in the histological slices. In untreated control wounds, the mean neoepidermal thickness was 22.50 µm with a range of 12.52–39.5 µm. Epidermal thickness had increased from treatment with the non-crosslinked biomatrix at a mean thickness of 31.01 µm (range 23.03–61.45 µm; $p = 0.0015$) and from the crosslinked biomatrix at a mean of 43.12 µm (range of 25.78–173.09 µm; $p < 0.0001$), as shown in Figure 3. Thus, the histological examination revealed higher epidermal thickness in the crosslinked biomatrix-derived neoepidermis. Rete ridge-like formations were visible in both treatment groups (Figure 1).

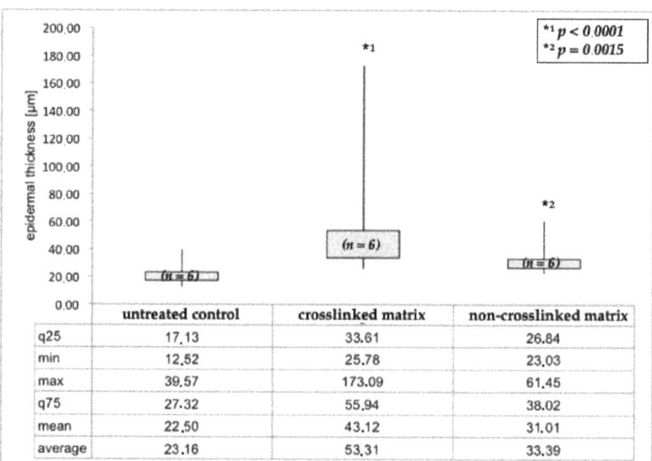

	untreated control	crosslinked matrix	non-crosslinked matrix
q25	17.13	33.61	26.84
min	12.52	25.78	23.03
max	39.57	173.09	61.45
q75	27.32	55.94	38.02
mean	22.50	43.12	31.01
average	23.16	53.31	33.39

Figure 3. Epidermal thickness of untreated wounds versus crosslinked and non-crosslinked biomatrices-derived neoepidermis in three different sections per sample with an interval of 100 µm in micrometers. Histologic examination verified highest epidermal thickness in the crosslinked biomatrix-treated wounds. * Statistically significant results.

In comparison, the mean epidermal cell count for untreated controls amounted to 41,00 cells/5 mm^2 (range of 22.00–68.00 cells/5 mm^2). Higher values of cell density were seen in the crosslinked biomatrix-treated groups with 69.50 cells/5 mm^2 (range of 57.00–107.00 cells/5 mm^2, $p = 0.0089$), whereas in the non-crosslinked biomatrix-treated wounds the highest values for mean neoepidermal cell count with an average of 84.50 cells/5 mm^2 (range of 79.00–106.00 cells/5 mm^2, $p = 0.0002$) could be observed. The templates were fully integrated into the host tissue with the colonization of epidermal minipig cells within all sections at statistically significant levels (Figure 4).

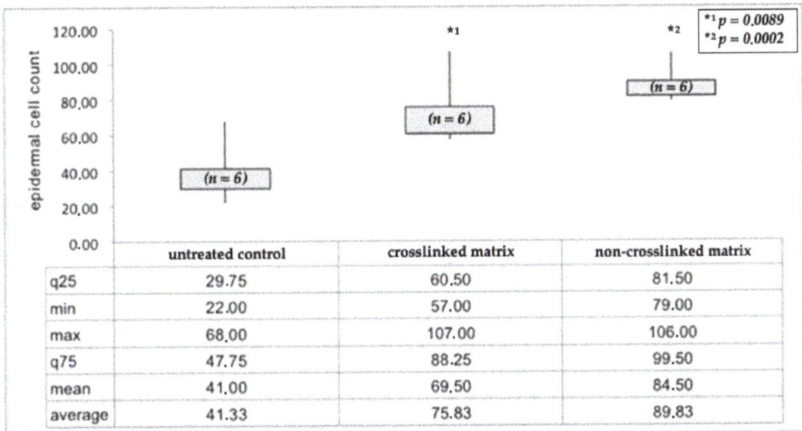

	untreated control	crosslinked matrix	non-crosslinked matrix
q25	29.75	60.50	81.50
min	22.00	57.00	79.00
max	68.00	107.00	106.00
q75	47.75	88.25	99.50
mean	41.00	69.50	84.50
average	41.33	75.83	89.83

Figure 4. Neoepidermal cell density in untreated, the crosslinked and non-crosslinked biomatrices-derived neoepidermis within a section of 100 μm width taken repetitively in three different sections. Histologic examination demonstrated the densest colonization of neoepidermal cells in the non-crosslinked biomatrix-treated wounds. * Statistically significant results.

To sum up, treatment groups showed overall improved wound healing with accelerated wound closure, thicker neoepidermal tissue and increased epidermal cell density in comparison to untreated wounds.

4. Discussion

Full-thickness skin injuries, such as deep burn wounds, generally require surgical intervention in addition to sufficient debridement. In medical care, autologous skin grafting is still the gold standard of dermal replacement [15–17]. However, when harvesting a skin graft, a new wound surface will arise and might result in hypertrophic scarring, severe pain, unsatisfactory cosmetic and functional complications or even additional chronic wounds in patients with microcirculation disorders [18]. In burn patients, the harvesting of a graft leads to a larger wound area with an increased risk of fluid loss, infection and temperature loss. Moreover, in extensive burns, the donor region is limited, so the overall aim in plastic, reconstructive and burn surgery remains the use of skin analogues similar to natural skin. There are various manufactured skin substitutes, but at present, there is still no commercially available skin substitute that can restore all necessary characteristics of the native skin [19]. Beginning in the early 1980s, tissue engineering developed some skin replacement products with promising results with deep dermal burn wounds and for use in reconstructive surgery [20–23], including the analyzed crosslinked and non-crosslinked biomatrices. Dermal substitutes such as these are mainly acellular 3-dimensional scaffolds and of allogenic, xenogenic or synthetic origin, which nourish the epidermal layer or an epidermal graft (i.e., skin grafts). Both analyzed products contain collagen, the key component of the human skin. Cell migration takes place along the collagen fibers in an organized manner.

The non-crosslinked biomatrix is used for dermal regeneration with promising results comparable to pure full-thickness grafts but with marginally slower regeneration compared with split-thickness skin grafting only [24]. Even in wet conditions it retains its stability and elasticity [25] and improves the biomechanical properties of the regenerating tissue due to the elastin component [26]. It reduces wound contraction and scar formation, so the reconstructed skin is stress-resistant sooner and patients' rehabilitation is faster, shortening their hospital stay and sophisticating long-term results [27,28].

The crosslinked biomatrix also serves as a dermal regeneration template. The dermal matrix with its open pore structure promotes the ingrowth of fibroblasts and endothelial cells [29]. Its silicone layer functions as a temporary barrier preventing fluid loss and protects against mechanical influences and bacterial contamination [20,25,30–33]. Usually, the silastic sheet may be removed 15 to 28 days after placement [29]. In our study, it was left on the wound for seven days because we already observed wound healing and did not want to integrate the silicone into the wound and compromise epithelialization. Klein et al. already showed that deep facial burns tolerate the removal of the silicone layer after seven to ten days without difficulties [34]. By applying the crosslinked biomatrix complex, skin defects, even with exposed anatomical structures, can be managed and this process might even reduce the necessity for reconstructive techniques, such as local or free flap transplantation [35]. Nevertheless, artificial skin replacement products have disadvantages, including their relative expense, the risk of infection complications and, if the treatment is insufficient, patients will suffer fragile wounds and consequent surgical revisions [36].

The crosslinked and non-crosslinked biomatrices are usually applied together with split-thickness skin grafting, as Böttcher–Haberzeth et al. did in rats [5]. They demonstrated that the crosslinked and non-crosslinked biomatrices offer promising approaches for one-step closure with a neonatal rat epidermis. The crosslinked biomatrix-derived neodermis was thicker than in the non-crosslinked biomatrix-treated wounds, and cell density was higher in the non-crosslinked biomatrix-derived neodermis, same as in our study concerning the neoepidermis. The analyzation of the data with a particular focus on neoepidermal formation, even though the products are intended for use in dermal replacement, represents a further discrepancy in the recommended handling. Nevertheless, in the present study, the two biomaterials were investigated as a dermal replacement and a single promotor for epithelialization at the same time without additional skin grafting, and, therefore, no donor site morbidity. Both treatment groups showed significantly faster reepthelialization than the controls, which suggests the presumption that the dermal substitutes offer a better surface for keratinocytes to move from the wound edges. Furthermore, the histologic examination demonstrated rete ridge-like formations especially in the non-crosslinked biomatrix-treated wounds. These are advantageous because the surface area of the epidermis is increased at the dermoepidermal junction, and the cells of the epidermis receive their nutrients through improved contact with the blood vessels in the dermis. In addition, rete ridges provide a niche for stem cells, protected from environmental stress with epidermal shearing, such as shear forces. With this, support of skin regeneration could be observed through the application of the dermis replacement products and can even reduce the need for local or free flap-plastic defect coverage [35].

However, only minor wounds reepithelialize without the need for transplantation [37] as also shown by Fulchignoni et al. They recently published a clinical study, comparing the use of the non-crosslinked biomatrix with or without grafting in the treatment of fingertip tissue loss [38]. So, in the current study, it would have been an interesting point to analyze wounds of increasing sizes. Moreover, additional data on the functionality of the skin and vascularization would have been further interesting aspects.

This led us to look again at this issue. The experimental setup chosen in our study turned out to be a reliable option as a wound-healing model [9]. In general, porcine skin shows recognizable similarities to human skin in terms of its anatomical composition and physiological behavior, as described extensively in the literature [39–41].

5. Conclusions

In conclusion, this trial showed comparable biological behavior with successful wound closure using crosslinked and non-crosslinked biomatrices in full-thickness wounds. Our data presented a positive influence on the epidermal regeneration and support of epithelialization with the chosen dermis substitutes even without additional skin transplantation and, thus, without additional donor site morbidity. Therefore, it can be stated that the single biomatrix application might be used in small wounds, which needs to be investigated further in a clinical setting to determine the size and depths of a suitable wound bed. Nevertheless, currently available products cannot solely achieve wound healing that is equal to or superior to autologous tissue. Thus, the overarching aim still is the development of an innovative skin substitute to manage surface reconstruction without additional skin grafting.

Author Contributions: Conceptualization and methodology, W.E., M.H. and A.R.-S.; software, J.-O.B. and M.H.; formal analysis, J.-O.B. and M.D.; validation, investigation, and data curation, W.E., J.-O.B. and M.H.; resources, A.R.-S. and A.D.; writing—original draft preparation, W.E. and M.D.; writing—review and editing, W.E.; visualization, M.H. and M.D.; supervision and project administration, A.R.-S. and A.D.; funding acquisition, W.E.; All authors have read and agreed to the published version of the manuscript.

Funding: The first and corresponding author received financial support from the research promotion program TÜFF of the local university (application number 2459-0-0). Further we acknowledge support by the Open Access Publishing Fund of the University of Tübingen in Germany.

Institutional Review Board Statement: This study was conducted in accordance with the Declaration of Helsinki. Animals were treated according to the German Law on the Protection of Animals, and the study was performed with permission from the local animal welfare committee (approval code AT 1/12) of the University of Tübingen in Germany.

Acknowledgments: There are no further contributions that do not justify authorship but need acknowledgment and no acknowledgment of technical help.

Conflicts of Interest: The authors disclose no conflict of interest.

References

1. Auger, F.A.; Berthod, F.; Moulin, V.; Pouliot, R.; Germain, L. Tissue-engineered skin substitutes: From in vitro constructs to in vivo applications. *Biotechnol. Appl. Biochem.* **2004**, *39*, 263–275. [CrossRef] [PubMed]
2. Dieckmann, C.; Renner, R.; Milkova, L.; Simon, J.C. Regenerative medicine in dermatology: Biomaterials, tissue engineering, stem cells, gene transfer and beyond. *Exp. Derm.* **2010**, *19*, 697–706. [CrossRef] [PubMed]
3. van der Veen, V.C.; van der Wal, M.B.; van Leeuwen, M.C.; Ulrich, M.M.; Middelkoop, E. Biological background of dermal substitutes. *Burns* **2010**, *36*, 305–321. [CrossRef] [PubMed]
4. Bottcher-Haberzeth, S.; Biedermann, T.; Reichmann, E. Tissue engineering of skin. *Burns* **2010**, *36*, 450–460. [CrossRef]
5. Bottcher-Haberzeth, S.; Biedermann, T.; Schiestl, C.; Hartmann-Fritsch, F.; Schneider, J.; Reichmann, E.; Meuli, M. Matriderm(R) 1 mm versus Integra(R) Single Layer 1.3 mm for one-step closure of full thickness skin defects: A comparative experimental study in rats. *Pediatr. Surg. Int.* **2012**, *28*, 171–177. [CrossRef]
6. Hodgkinson, T.; Bayat, A. Dermal substitute-assisted healing: Enhancing stem cell therapy with novel biomaterial design. *Arch. Derm. Res.* **2011**, *303*, 301–315. [CrossRef]
7. Garcia-Gareta, E.; Ravindran, N.; Sharma, V.; Samizadeh, S.; Dye, J.F. A novel multiparameter in vitro model of three-dimensional cell ingress into scaffolds for dermal reconstruction to predict in vivo outcome. *Biores. Open Access* **2013**, *2*, 412–420. [CrossRef]
8. Qvist, M.H.; Hoeck, U.; Kreilgaard, B.; Madsen, F.; Frokjaer, S. Evaluation of Gottingen minipig skin for transdermal in vitro permeation studies. *Eur. J. Pharm. Sci.* **2000**, *11*, 59–68. [CrossRef]
9. Sullivan, T.P.; Eaglstein, W.H.; Davis, S.C.; Mertz, P. The pig as a model for human wound healing. *Wound Repair Regen.* **2001**, *9*, 66–76. [CrossRef]
10. Held, M.; Rahmanian-Schwarz, A.; Schiefer, J.; Rath, R.; Werner, J.O.; Rahmanian, S.; Schaller, H.E.; Petersen, W. A Novel Collagen-Gelatin Scaffold for the Treatment of Deep Dermal Wounds-An Evaluation in a Minipig Model. *Derm. Surg.* **2016**, *42*, 751–756. [CrossRef]
11. Petersen, W.; Rahmanian-Schwarz, A.; Werner, J.O.; Schiefer, J.; Rothenberger, J.; Hubner, G.; Schaller, H.E.; Held, M. The use of collagen-based matrices in the treatment of full-thickness wounds. *Burns* **2016**, *42*, 1257–1264. [CrossRef] [PubMed]

12. Schiefer, J.L.; Held, M.; Fuchs, P.C.; Demir, E.; Ploger, F.; Schaller, H.E.; Rahmanian-Schwarz, A. Growth Differentiation Factor 5 Accelerates Wound Closure and Improves Skin Quality During Repair of Full-Thickness Skin Defects. *Adv. Ski. Wound Care* **2017**, *30*, 223–229. [CrossRef]
13. Schiefer, J.L.; Rath, R.; Held, M.; Petersen, W.; Werner, J.O.; Schaller, H.E.; Rahmanian-Schwarz, A. Frequent Application of the New Gelatin-Collagen Nonwoven Accelerates Wound Healing. *Adv. Ski. Wound Care* **2016**, *29*, 73–78. [CrossRef] [PubMed]
14. Fu, X.; Fang, L.; Li, H.; Li, X.; Cheng, B.; Sheng, Z. Adipose tissue extract enhances skin wound healing. *Wound Repair. Regen.* **2007**, *15*, 540–548. [CrossRef] [PubMed]
15. Andreassi, A.; Bilenchi, R.; Biagioli, M.; D'Aniello, C. Classification and pathophysiology of skin grafts. *Clin. Derm.* **2005**, *23*, 332–337. [CrossRef]
16. Stanton, R.A.; Billmire, D.A. Skin resurfacing for the burned patient. *Clin. Plast. Surg.* **2002**, *29*, 29–51. [CrossRef]
17. Supp, D.M.; Boyce, S.T. Engineered skin substitutes: Practices and potentials. *Clin. Derm.* **2005**, *23*, 403–412. [CrossRef]
18. Papini, R. Management of burn injuries of various depths. *BMJ* **2004**, *329*, 158–160. [CrossRef]
19. Murphy, P.S.; Evans, G.R. Advances in wound healing: A review of current wound healing products. *Plast. Surg. Int.* **2012**, *2012*, 190436. [CrossRef]
20. Burke, J.F.; Yannas, I.V.; Quinby, W.C., Jr.; Bondoc, C.C.; Jung, W.K. Successful use of a physiologically acceptable artificial skin in the treatment of extensive burn injury. *Ann. Surg.* **1981**, *194*, 413–428. [CrossRef]
21. Heimbach, D.; Luterman, A.; Burke, J.; Cram, A.; Herndon, D.; Hunt, J.; Jordan, M.; McManus, W.; Solem, L.; Warden, G.; et al. Artificial dermis for major burns. A multi-center randomized clinical trial. *Ann. Surg.* **1988**, *208*, 313–320. [CrossRef] [PubMed]
22. Heimbach, D.M.; Warden, G.D.; Luterman, A.; Jordan, M.H.; Ozobia, N.; Ryan, C.M.; Voigt, D.W.; Hickerson, W.L.; Saffle, J.R.; DeClement, F.A.; et al. Multicenter postapproval clinical trial of Integra dermal regeneration template for burn treatment. *J. Burn. Care Rehabil.* **2003**, *24*, 42–48. [CrossRef] [PubMed]
23. Heitland, A.; Piatkowski, A.; Noah, E.M.; Pallua, N. Update on the use of collagen/glycosaminoglycate skin substitute-six years of experiences with artificial skin in 15 German burn centers. *Burns* **2004**, *30*, 471–475. [CrossRef] [PubMed]
24. Kolokythas, P.; Aust, M.C.; Vogt, P.M.; Paulsen, F. [Dermal subsitute with the collagen-elastin matrix Matriderm in burn injuries: A comprehensive review]. *Handchir Mikrochir Plast Chir* **2008**, *40*, 367–371. [CrossRef]
25. Vogt, P.M.; Kolokythas, P.; Niederbichler, A.; Knobloch, K.; Reimers, K.; Choi, C.Y. Innovative wound therapy and skin substitutes for burns. *Der Chir. Z. Fur Alle Geb. Der Oper. Medizen* **2007**, *78*, 335–342. [CrossRef]
26. Ryssel, H.; Gazyakan, E.; Germann, G.; Ohlbauer, M. The use of MatriDerm in early excision and simultaneous autologous skin grafting in burns—A pilot study. *Burns* **2008**, *34*, 93–97. [CrossRef]
27. Haslik, W.; Kamolz, L.P.; Manna, F.; Hladik, M.; Rath, T.; Frey, M. Management of full-thickness skin defects in the hand and wrist region: First long-term experiences with the dermal matrix Matriderm. *J. Plast. Reconstr. Aesthet. Surg.* **2010**, *63*, 360–364. [CrossRef]
28. Haslik, W.; Kamolz, L.P.; Nathschlager, G.; Andel, H.; Meissl, G.; Frey, M. First experiences with the collagen-elastin matrix Matriderm as a dermal substitute in severe burn injuries of the hand. *Burns* **2007**, *33*, 364–368. [CrossRef]
29. Shores, J.T.; Gabriel, A.; Gupta, S. Skin substitutes and alternatives: A review. *Adv. Ski. Wound Care* **2007**, *20*, 493–508. [CrossRef]
30. Atiyeh, B.S.; Hayek, S.N.; Gunn, S.W. New technologies for burn wound closure and healing–review of the literature. *Burns* **2005**, *31*, 944–956. [CrossRef]
31. Bello, Y.M.; Falabella, A.F.; Eaglstein, W.H. Tissue-engineered skin. Current status in wound healing. *Am. J. Clin. Derm.* **2001**, *2*, 305–313. [CrossRef] [PubMed]
32. Grzesiak, J.J.; Pierschbacher, M.D.; Amodeo, M.F.; Malaney, T.I.; Glass, J.R. Enhancement of cell interactions with collagen/glycosaminoglycan matrices by RGD derivatization. *Biomaterials* **1997**, *18*, 1625–1632. [CrossRef]
33. Murray, R.C.; Gordin, E.A.; Saigal, K.; Leventhal, D.; Krein, H.; Heffelfinger, R.N. Reconstruction of the radial forearm free flap donor site using integra artificial dermis. *Microsurgery* **2011**, *31*, 104–108. [CrossRef] [PubMed]
34. Klein, M.B.; Engrav, L.H.; Holmes, J.H.; Friedrich, J.B.; Costa, B.A.; Honari, S.; Gibran, N.S. Management of facial burns with a collagen/glycosaminoglycan skin substitute-prospective experience with 12 consecutive patients with large, deep facial burns. *Burns* **2005**, *31*, 257–261. [CrossRef]
35. Helgeson, M.D.; Potter, B.K.; Evans, K.N.; Shawen, S.B. Bioartificial dermal substitute: A preliminary report on its use for the management of complex combat-related soft tissue wounds. *J. Orthop Trauma* **2007**, *21*, 394–399. [CrossRef]
36. Chou, T.D.; Chen, S.L.; Lee, T.W.; Chen, S.G.; Cheng, T.Y.; Lee, C.H.; Chen, T.M.; Wang, H.J. Reconstruction of burn scar of the upper extremities with artificial skin. *Plast. Reconstr. Surg.* **2001**, *108*, 378–384. [CrossRef]
37. Wisser, D.; Steffes, J. Skin replacement with a collagen based dermal substitute, autologous keratinocytes and fibroblasts in burn trauma. *Burns* **2003**, *29*, 375–380. [CrossRef]
38. Fulchignoni, C.; Rocchi, L.; Cauteruccio, M.; Merendi, G. Matriderm dermal substitute in the treatment of post traumatic hand's fingertip tissue loss. *J. Cosmet. Dermatol.* **2022**, *21*, 750–757. [CrossRef] [PubMed]
39. Kwak, M.; Son, D.; Kim, J.; Han, K. Static Langer's line and wound contraction rates according to anatomical regions in a porcine model. *Wound Repair Regen.* **2014**, *22*, 678–682. [CrossRef]

40. Riccobono, D.; Forcheron, F.; Agay, D.; Scherthan, H.; Meineke, V.; Drouet, M. Transient gene therapy to treat cutaneous radiation syndrome: Development in a minipig model. *Health Phys.* **2014**, *106*, 713–719. [CrossRef]
41. Rothenberger, J.; Held, M.; Jaminet, P.; Schiefer, J.; Petersen, W.; Schaller, H.E.; Rahmanian-Schwarz, A. Development of an animal frostbite injury model using the Goettingen-Minipig. *Burns* **2014**, *40*, 268–273. [CrossRef] [PubMed]

MDPI
St. Alban-Anlage 66
4052 Basel
Switzerland
Tel. +41 61 683 77 34
Fax +41 61 302 89 18
www.mdpi.com

Applied Sciences Editorial Office
E-mail: applsci@mdpi.com
www.mdpi.com/journal/applsci

www.ingramcontent.com/pod-product-compliance
Lightning Source LLC
LaVergne TN
LVHW070610170726
843515LV00004B/33

* 9 7 8 3 0 3 6 5 5 3 1 7 7 *